Wisconsin Waters

The Tyler Forks River plummets over Brownstone Falls to merge with the Bad River in Copper Falls State Park.

WISCONSIN
WATERS

The Ancient History of Lakes, Rivers, and Waterfalls

Scott Spoolman

WISCONSIN HISTORICAL SOCIETY PRESS

Published by the Wisconsin Historical Society Press
Publishers since 1855

The Wisconsin Historical Society helps people connect to the past by collecting, preserving, and sharing stories. Founded in 1846, the Society is one of the nation's finest historical institutions. *Join the Wisconsin Historical Society:* wisconsinhistory.org/membership

Photographs are by Scott Spoolman unless otherwise noted.
The front cover photo shows a view of the Wisconsin River as it flows through the Wisconsin Dells.

Printed in Canada
Designed by Mayfly Design

26 25 24 23 22   1 2 3 4 5

Library of Congress Cataloging-in-Publication Data
Names: Spoolman, Scott, author.
Title: Wisconsin waters : the ancient history of lakes, rivers, and waterfalls / Scott Spoolman.
Description: Madison : Wisconsin Historical Society Press, 2022. | Includes bibliographical
    references and index.
Identifiers: LCCN 2021060251 (print) | LCCN 2021060252 (e-book) | ISBN 9780870209949
    (paperback) | ISBN 9780870209963 (epub)
Subjects: LCSH: Hydrology—Wisconsin. | Bodies of water—Wisconsin.
Classification: LCC GB705.W6 S76 2022 (print) | LCC GB705.W6 (e-book) |
    DDC 551.4809775—dc23/eng20220406
LC record available at https://lccn.loc.gov/2021060251
LC e-book record available at https://lccn.loc.gov/2021060252

This one is for John, who handed me my first canoe paddle. He led us to many adventures on the waters and into the woods. He loved and embraced the natural world as much as anyone I've ever known.

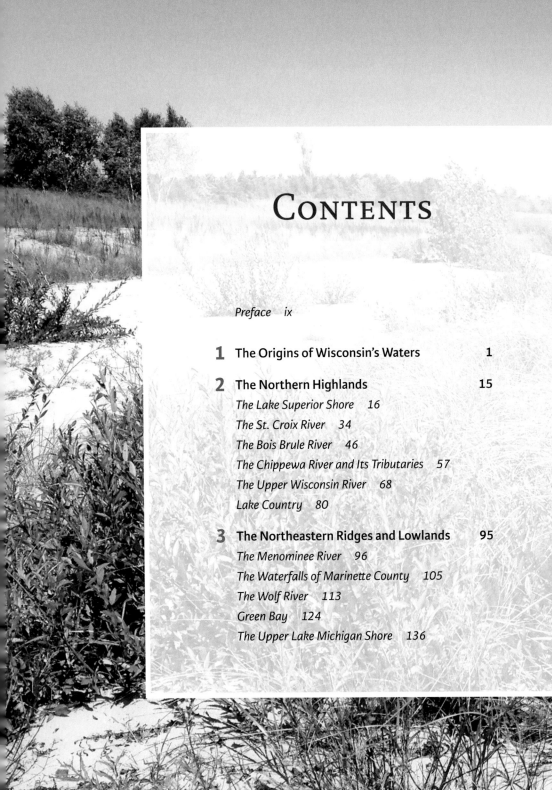

CONTENTS

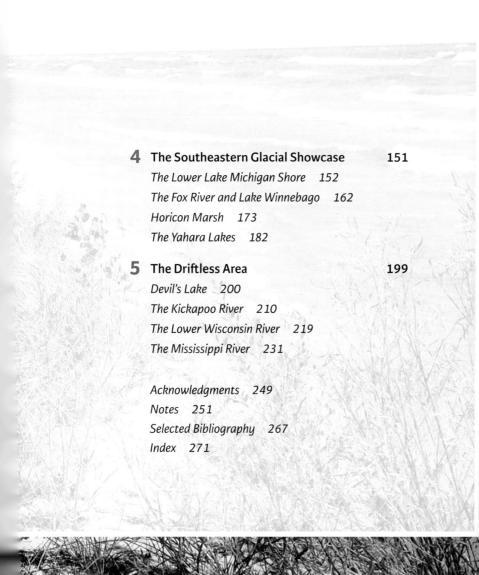

Willow Falls at Willow
River State Park

PREFACE

Tucked into a corner of the upper Midwest formed by two of the Great Lakes and bordered on its west side by North America's largest river, Wisconsin is framed by waterways and rich in various water features. Within its borders lie 15,000 lakes, most of them formed by glaciers some 10,000 years ago. The state is laced with more than 12,600 rivers and streams, including 2,700 trout streams and three designated as National Wild and Scenic Rivers. These streams and rivers trickle or surge over dozens of waterfalls that mesmerize thousands of visitors every year. Wisconsin's economy depends largely on its waters, which support its major industries, including manufacturing, dairy farming, paper production, commercial fishing, cranberry production, and tourism. And water has done much to shape the history of the state.

Each of Wisconsin's waterways has a story to tell. Some stories begin hundreds of millions of years ago with the formation of underlying bedrock. Others begin thousands of years ago with the advance and retreat of glaciers that molded landscapes. Still others have to do with human stories of exploration and settlement of the land and the development of societies and economies. These stories together create a complex tapestry, woven together in time and space by the flow of Wisconsin's waters.

While stories of how people have used Wisconsin's waters could fill several volumes, this book looks back through the distant past to the ancient origins of Wisconsin's lakes, rivers, waterfalls, and wetlands, beginning with geology—how the bedrock of the state was formed. This view of the past also takes in the natural history—the old and continuing story of evolving plant and animal communities—that has shaped, and been shaped by, water since the geologic stage for that story was built. And it includes some of the early tales of how

The lagoon at Big Bay State Park on Madeline Island, with kayakers in the distance

humans lived along Wisconsin's waterways before people began to engineer and dramatically change them for their own uses.

It was not easy for me to choose which of the state's water features to cover in this book, given the thousands of options available. Some choices were obvious. The Mighty Mississippi and the two Great Lakes' shores top that list. As for the myriad lakes, rivers, waterfalls, and wetlands, I strove to choose a limited number that would represent the others in terms of their types and ancient histories. No doubt, I've left out many readers' favorites. If there were no space limitations, I would gladly cover them all.

The stories are organized into four geographic sections of the state, with an assortment of waterways in each. Chapter 1, on the origins of Wisconsin's waters, includes an overview of the state's ancient past with details on important historical processes such as bedrock formation, glacial dynamics, and the development of glacial landforms. It is not necessary to read the other chapters in any certain order, but I do recommend reading Chapter 1 first to get a refresher on these key topics and to refer back to it when reading the rest of the book. Likewise, the short introductions to Chapters 2 through 5 provide a few more details as reference material on the deep histories of their respective quarters of the state.

Finally, each section of Chapters 2 through 5 includes a travel guide—a suggested route for exploring the waterway—with notes on what you may see as evidence of its geologic and natural history. In these guides, the distances mentioned are approximate. Including a map of each route was not feasible, so these descriptions are intended to be supplemented by a good set of maps. With that, I hope you enjoy your explorations of Wisconsin's waters.

1.1 Upper Falls of Amnicon
Falls State Park

1

The Origins of
Wisconsin's Waters

Where did all the water on our planet come from? The long-running debate on this question continues to this day. Scientists widely agree that hydrogen was generated by the Big Bang 13.8 billion years ago and that, beginning a few billion years later, oxygen was manufactured by stars and spewed across the galaxy by supernova explosions. Hydrogen and oxygen then comingled to make water molecules that are thought to be widespread across the galaxy. The question is, how did oceans full of those H_2O molecules end up on Planet Earth? One long-held view is that Earth's water was deposited by comets and/or asteroids colliding with the planet sometime after its molten crust cooled enough to retain water. Another more recently explored theory is that water molecules existed in glass crystals contained in rocks deep in the crust and were somehow released early in the planet's history.[1]

Whatever its origin, water exists on our planet in higher concentrations than on any other body in the solar system, as far as we know. Earth holds some 366 trillion gallons of H_2O. If we take away the salty oceans, the ice at the planet's poles and in the mountain glaciers, and the water deep underground in untapped aquifers, we are left with just 0.024 percent of all the Earth's water. That is the amount available for use by all the planet's freshwater ecosystems, all the plants and animals they support, and all the world's 7.9 billion people.[2]

Of that total (but seemingly tiny) amount of freshwater, an even smaller fraction flows in the streams, rivers, lakes, wetlands, and aquifers of Wisconsin. When we think of the state's many waterfalls gushing millions of gallons of water every hour of every day, or of the hundreds of deep lakes strewn across much of Wisconsin's landscape, the amount of water within the state's borders at any given time seems infinite. It is becoming increasingly important to realize, though, that the total amount is not unlimited. The amount of freshwater within any town, county, state, country—and the total amount available on Earth—is finite. It is pollutable and exhaustible, and because we depend on it for our lives and livelihoods, we need to take care of our supply and use it wisely.

What might help us to understand the finite and precious nature of freshwater is to learn about how our present-day waterscapes were formed. The focus of this book is the enormity of the processes that have shaped our land and water features, along with the unimaginably long periods of time required by those processes. For that, we first need to consider the overall geological and natural history of Wisconsin.

An Overview of Wisconsin's Ancient Past

About 3.5 billion years ago, during the early Precambrian era (Figure 1.2), the oldest known rocks were formed in the region that would include Wisconsin. Magma—hot, fluid rock from the earth's mantle—flowed into fissures and chambers underground and erupted onto the surface, forming igneous rock such as granite (created underground) and basalt (created on the surface). Geologic forces fractured and buried some of this rock miles below the surface where it was metamorphosed—deformed and reconfigured by unimaginable heat and pressure—to become metamorphic rock, part of the foundation of the continent.

During the ensuing eons, shallow seas invaded and retreated from the region several times, each time covering all or most of Wisconsin and depositing layers of sand and other sediments on the sea floor. Over millions of years, the increasing weight of these layers, along with the heat of compression and certain chemical reactions, converted these sediments into sedimentary rock—the third major rock type after igneous and metamorphic rock. Whenever seas receded, the barren, wind-whipped land experienced vast erosion, usually for many millions of years, which reversed the process of deposition by deforming or dismantling ancient rock layers.

Another major force that shaped Wisconsin's bedrock was plate tectonics—the movement of the earth's crustal plates resulting in continental collisions, volcanic activity, and mountain building. Over a period of about 75 million

AGE	PERIOD		YEARS AGO
Cenozoic	Quaternary	Holocene epoch	12,000
		Pleistocene epoch	2.6 mya
	Tertiary		65 mya
Mesozoic	Cretaceous		145 mya
	Jurassic		208 mya
	Triassic		248 mya
Paleozoic	Permian		286 mya
	Pennsylvanian		320 mya
	Mississippian		360 mya
	Devonian		417 mya
	Silurian		444 mya
	Ordovician		495 mya
	Cambrian		545 mya
Precambrian	Proterozoic eon		2,500 mya
	Archean eon		

1.2 The geologic timetable. Below the Quaternary Period, units of time are listed as millions of years ago, or MYA. ADAPTED FROM DOTT AND ATTIG, *ROADSIDE GEOLOGY OF WISCONSIN*

years, one continental collision heaved up the Penokean Mountains, which stretched across parts of present-day Minnesota, northern Wisconsin and Michigan, and southern Ontario, occupying a footprint similar to that of today's Appalachian Mountains. At the time—around 1.85 billion years ago—due to plate tectonics, Wisconsin was south of the equator, and the Penokeans were tropical desert mountains. Over millennia, they would be dismantled by erosion, providing enormous volumes of sand and gravel to be shaped and reshaped on the ever-changing landscape by wind, flowing water, and ancient seas invading and retreating.

As the Penokeans eroded, southern Wisconsin, too, saw its share of continental collisions, volcanic activity, mountain building, and submersion by ancient seas. Between 1.7 and 1.6 billion years ago, deep layers of sandstone were metamorphosed into the hard rock quartzite, an important component of bedrock in several areas of the state.

By about 1.1 billion years ago, ancient seas had receded, and erosion was to become the primary force shaping the land for the next 500 million years or more. At this time, Wisconsin was still located near the equator, far from any coast, and its mountains were all but gone—the region being primarily a rolling tropical desert.[3] But its peace and quiet did not last, for deep in the earth's mantle, around 1.1 billion years ago, a gigantic plume of magma rose into the crust to begin a major event known as the Midcontinent Rift. The magma plume torched, melted, and churned rocks and fractured the surface in an arcing pattern that framed what would become Wisconsin. The top of the arc was the area that is now Lake Superior, and the two legs stretched away to the southwest and southeast (Figure 1.3).

The rift threatened to split the continent in two, but then, for unknown reasons, the rifting process halted. During the 25 million years between the opening and closing of the rift, however, magma intermittently erupted and flowed across a long, broad area centered on the crevice. As this spurting lava cooled, it became basalt—a fine-grained, dark-colored, very dense type of rock that now underlies much of northwestern Wisconsin. In the central rift zone, the land slowly subsided, or sank, due to the steadily increasing mass of basalt. This subsidence created a syncline—a huge trough-shaped depression lying across the area now occupied by Lake Superior. Geologists refer to this rifting period as the Keweenawan episode—named for Michigan's Keweenaw Peninsula, which juts into the Lake Superior basin.

1.3 The Midcontinent Rift and the Wisconsin Dome. MAPPING SPECIALISTS, LTD., FITCHBURG, WI

The rift possibly gave rise, quite literally, to another land feature important to Wisconsin's ancient history, the Wisconsin Dome. Some geologists think the gentle uplifting of north central Wisconsin was a process that complemented the subsidence of the land within the rift area, something like what can happen when you push down on two parts of a pillow and the spot between your hands pops up. Now the land in northern Wisconsin slopes gently on all sides from the central Wisconsin Dome, which has played an important role in the shaping of the state's land and waterscapes.[4]

At the end of Precambrian time, around 545 million years ago, North America was a tropical plain covered by a thin layer of sand that was moved around by winds and rivers. Then came the Paleozoic era, when another series of seas invaded the region. The first two of these were Cambrian seas fed by streams washing sand out of the Penokean highlands to form layers of sandstone that today undergird much of Wisconsin. These porous layers of stone store vast amounts of water in aquifers that serve as drinking water supplies for most of the state's population. Wisconsin's groundwater supply is estimated to be around 1.2 trillion gallons—enough to submerge the entire state under 100 feet of water were it all to be pumped to the surface.[5]

During the succeeding geologic periods, especially the Ordovician and Silurian, life in the Paleozoic seas evolved rapidly. Eventually, Wisconsin's oceans teemed with life. Corals built some of the earliest reefs on the planet around the Lake Michigan Basin, including southeastern Wisconsin. For 300 million years, the wastes and remains of ancient plants and animals accumulated on sea floors. Geologic processes converted those remains to limestone, which was in turn converted to dolomite, a related but harder rock that once covered all of Wisconsin in layers up to 600 feet thick. When the last of these inland seas withdrew from the continent, erosion became the defining geologic force. Over hundreds of millions of years, wind and flowing water removed most of those dolomite layers across the state, beginning with the higher areas such as the Wisconsin Dome. This long period of erosion erased the fossil record along with the dolomite, so we cannot be sure, for example, whether dinosaurs ever roamed the state (although chances are that they did). Ordovician and Silurian dolomite still form resistant bedrock in some areas of the state, most notably in the east.

The next big event in the shaping of Wisconsin's land and waterscapes was the coming of the Pleistocene epoch—the Ice Age and the time of the glaciers. By 2.5 million years ago, North America had moved to its current location and enjoyed a mild, moist climate. The land hosted a rich diversity of prairies, forests, wetlands, and animals, including mammoths, giant ground sloths, camels, saber-toothed cats, and the ancestors of today's rhinoceroses, horses, dogs, and cats.[6]

Then the climate began to cool due to a complicated cycle involving long-term periodic changes in Earth's orbit around the sun and in the tilt of the planet's axis. In parts of the northern hemisphere, the climate cooled to the point where winter snow piled up and stayed through the warm months of the year. Each year, more snow remained, and as this supply of snow grew larger and deeper, the weight of overlying snow layers squeezed the air out of the layers beneath them. This process converted the lower layers to ice that accumulated, becoming hundreds of feet thick in certain areas. When such a mass of ice grows thicker than about 100 feet, gravity begins to pull it down any underlying slopes. As more snow is added, the mass expands and spreads, and when it begins to move in this way, the ice mass has become a glacier.

This formation of glaciers happened on a massive scale during the Pleistocene. Over the next 2 to 2.5 million years, about 30 percent of the planet's land area would be covered several times with ice. (Today, ice covers about 10 percent of the earth's surface.) Thousands of years after the first advance of ice from the north, the climate warmed and the glaciers melted and retreated north to where they had first formed. Then the climate cooled and the cycle began again. Geologists estimate that this glaciation cycle occurred 10 or more times at roughly regular intervals over the past 2.5 million years.[7]

North America's largest glacier, the Laurentide Ice Sheet, named for Quebec's Laurentian Mountains near its center, originated around Hudson Bay, straight north of present-day Wisconsin. Time and time again, it spread in all directions, expanding along whatever course the lay of the land allowed. At the peak of its most recent advance, it extended northward to the Arctic Ocean and from the west coast to the east coast of Canada. It reached into the northern tier of the United States and dipped as far south as Illinois and Missouri.

That most recent advance—named the Wisconsin glaciation because the state contains some of the region's best representations of glacial effects on the land—occurred roughly between 100,000 and 11,000 years ago. It crossed what would be Wisconsin's northern border about 30,000 years ago and, within the next 7,000 years, covered all of northern and eastern Wisconsin. Scientists estimate that the innermost part of the ice sheet grew to thousands of feet thick, tapering to hundreds of feet at its edges. The weight of this mass of ice depressed the land, in some places by more than 300 feet.[8]

The glacier flowed in separate lobes, each resembling an enormous tongue of ice (Figure 1.4). In Wisconsin, all but one of these lobes are named for the lowlands through which they flowed. The Superior Lobe and its predecessors in previous glaciers slowly carved out the basin of Lake Superior. The Chippewa Lobe followed the contours of today's Chippewa River Valley. The Wisconsin Valley Lobe flowed in the river valley for which it is named. The Langlade Lobe, so named because it terminated in what later became Langlade County, flowed into the valley of the Wolf River. The Green Bay Lobe was guided by, and scooped out much of, the lowland that contains Green Bay, the Fox River, Lake Winnebago, and the Horicon Marsh. Finally, the Lake Michigan Lobe excavated the basin containing Lake Michigan and covered much of easternmost Wisconsin.

The southwest corner of the state was not touched by any of these six lobes, nor was it ever glaciated. That fact makes Wisconsin geologically unique among

1.4 The Wisconsin glaciation. The most recent glacier in Wisconsin was composed of six major lobes. Arrows indicate the general direction of flow for each lobe. WISCONSIN GEOLOGICAL AND NATURAL HISTORY SURVEY

all northern states. Years ago, geologists used the word *drift* to describe the sand, gravel, and boulders carried by glaciers and dropped on the land as the ice melted back. Wisconsin's Driftless Area, which includes all or parts of 22 counties, has no evidence of drift. Geologists conclude that the area was never ice-covered, even though glaciers went as far south as the Missouri and Ohio River Valleys. Thus, the area has never undergone the massive erosion that glaciers cause, which makes its landscapes and waterways distinctly different from those in the rest of Wisconsin.

The glaciers excavated the areas they covered. Every time the ice inched across the land, it carved bedrock to create stunning escarpments and promontories. It crushed softer layers of rock near the surface, picked up rock fragments, sand, gravel, and boulders by the ton, and moved it all forward. Each advance pushed the glacier to a line where it could go no farther, when ice at the front edge began melting as fast as the glacier could push more ice to that margin. The front wall of the glacier would then sit in one position for centuries while sand, gravel, and boulders were hauled by moving ice as on a giant conveyor belt to the margin, creating a long, broad zone of debris called a moraine. The Wisconsin glaciation put the finishing touches on the Great Lakes basins we see today. The Lake Superior Basin was gouged out to depths of more than 13,000 feet.

Another prominent landform shaped by advancing ice was the drumlin. These oblong hills number in the thousands across glaciated Wisconsin. Parts of the land surface under the most recent glacier were frozen hard, and other parts were relatively wet and soft. As the ice advanced, it picked up and pushed forward the softer material on the ground, while it rode up and over the frozen sections, molding them into teardrop-shaped hills. For that reason, drumlins are always oriented in the same direction as the flow of the glacier that formed them.

Even as they melted back when the climate warmed after each glaciation, the ice sheets affected the land and waterscapes dramatically. Water pooled and flowed on top of the retreating ice mass, making supraglacial lakes and streams. Under certain conditions, water flowed under the glacier in subglacial streams. Meltwater also flowed in copious sheets and streams off the top of the glacier.

Together, these flows of meltwater carried untold tons of clay, sand, and gravel. Some supraglacial streams bored vertical shafts in the ice and plunged from the glacier's top to its bottom, where they deposited their loads of debris, building cone-shaped mounds called kames. Some of these plummeting streams became subglacial streams, many of which, under high pressure caused by the weight of the ice mass, carved troughs called tunnel channels into the land under the ice. Where the streams gushed out from under the ice, they pushed loads of sand and gravel onto the ice-free land and, flowing in voluminous braided streams, sorted much of this debris and spread it out on fan-shaped expanses called outwash plains. As the subglacial streams eventually slowed, they dropped their debris onto the tunnel floors, forming long, sometimes sinuous deposits called eskers. Finally, the retreating ice often temporarily dammed large volumes of meltwater, and when the dams broke, massive torrents of ice water raged through the drainage routes, tearing away rock and carving exquisite landforms and river canyons.

By the time the ice left Wisconsin about 9,500 years ago, it had dropped vast fields of clay, sand, gravel, and boulders known as till. Out of this material, the glaciers and their meltwaters had built moraines, drumlins, kames, and eskers that now stood as ridges and hills, random arrangements of which are called hummocky terrain. The retreating glacier had also dropped great chunks of ice that were then buried under outwash. It took some of these ice masses decades and even centuries to melt, and in their places were left depressions called kettles, some of them miles wide, to be occupied by lakes, bogs, wetlands, or forest hollows.

Life gradually took hold on the postglacial landscape. As the glacial climate had encroached upon the region, forests and grasslands gradually died away or shifted their ranges, along with the animals that lived in those environs, and tundra conditions developed along the glacier's margin. This process reversed itself as the glacier melted away. Tundra slowly gave way to scattered open forests of spruce, fir, and jack pine. With continued warming, black spruce forests became established and spread north following the retreating ice. About 8,000 years ago, the postglacial warming accelerated, and conifer forests were edged out across much of the state by oak and elm, among other trees. White pine moved in from the east about 7,000 years ago and eventually dominated the state's forests. Prairies and savanna communities spread and flourished in many areas of the state.[9]

As plant communities changed, so did animal communities. At the peak of the most recent glaciation, so much of the earth's water was frozen that ocean levels were more than 300 feet lower than they are today. This was enough of a drop for the land under the Bering Strait, now separating Siberia from Alaska, to have emerged from under the sea. This strip of land, called Beringia, was around 1,000 miles wide, and for centuries after it was free of ice, it formed a bridge over which plants and animals could migrate from Asia to North America. Consequently, the postglacial tundra in the upper Midwest eventually hosted musk oxen, mastodons, mammoths, elk, caribou, white-tailed deer, wolves, giant beavers (larger than today's black bears), and other large mammals, or megafauna. One researcher has speculated that the upper Great Lakes, especially Lake Huron, may have hosted walruses and whales.[10]

The megafauna followed the retreating glacier, munching on the lichens, grasses, and shrubs of the tundra as they spread east and north across their new continent. Later, many such species disappeared in one of the earth's largest extinction bursts. (The causes of these extinctions are a subject of debate and still somewhat mysterious.) The forests and grasslands that replaced the tundra after thousands of years were home to small mammals including chipmunks, squirrels, mice, and foxes. Early residents of wetlands and lakes included loons, redwinged blackbirds, wood ducks, and muskrats, and plant communities hosted cattails, sedges, bog laurels, and other wetland plants.

By about 5,000 years ago, Wisconsin looked much as it did before Europeans immigrated in large numbers, beginning in the 1600s, changing the land for farming and other purposes and forcing out most of the many Native American people who had for centuries been part of the ecosystem. Mixed white pine and hardwood forests stood thick and deep across the northeastern half of the state, while the southwestern half hosted a mix of lush prairie and oak savanna. Wetlands sprawled across much of the state outside of the Driftless Area, and kettle lakes dotted the same region by the thousands. The entire state was laced with clear-flowing streams and rivers. The Driftless Area had been eroded and drained for so long by its arterial network of streams that no natural lakes remained. But even in that area, plenty of water continued to flow in the deep stream valleys.

It seems that humans have always been drawn to the water-rich paradise that is Wisconsin. Even the name of the state, derived from an Algonquian word, honors water. A number of explanations for the origin of the name have arisen over the years, and almost all of them have to do with water. The explanation now most accepted by historians is that the name comes from an Algonquian language spoken by Miami Indians who lived in the area when the French explorers Jacques Marquette and Louis Joliet traveled in Wisconsin in the summer of 1673. Their Miami guides referred to a river they explored using a word that Marquette recorded as "Meskousing," meaning "this stream meanders through something red." Scholars have concluded that the Miami guides were referring to today's Wisconsin River, which runs 430 miles across the state and, near the middle of its course, flows through the spectacular reddish sandstone gorges of the Wisconsin Dells.[11]

The first humans to come to Wisconsin were Paleo-Indians believed to have entered North America over a land bridge or possibly by boat. They were hunter-gatherers who entered Wisconsin about 12,000 years ago following the megafauna that roamed eastward and northward, crossing the continent along the margin of the glacier. The various Indigenous peoples who have lived in the

1.5 Peshtigo River rapids in Goodman County Park

region since that time gradually became less nomadic as their populations grew and they settled into particular regions, adapting to the varying landscapes and available water resources. During the time between the glacial retreat and the encroachment of European settlers, Indigenous people relied on hunting and gathering and developed gardening and food storage methods that allowed them to establish more permanent villages. They also took to the water in dugout canoes and, later, in far more versatile birch bark canoes, and they developed fishing practices to provide their diets with a major additional source of protein.

Beginning about 4000 BCE, people of the Late Archaic period began to make use of copper deposits along the south shore of Lake Superior. These early miners first collected copper nuggets that had worked their way to the surface, and later they found ways to mine shallow veins of copper. Members of what became known as the Old Copper culture produced copper spear points, knives, awls, needles, fishhooks, several other tools, and ornaments such as beads, bracelets, and headdress pieces. The Copper culture people are thought to be the first in the world to have domesticated dogs.[12] Similarly, Indigenous people in south-western Wisconsin found shallow deposits of lead and became the earliest lead miners in the area.

During the Middle Woodland period—2,300 to 1,500 years ago—people built villages along rivers in southern Wisconsin and on lakeshores in the north. Village-centered life is thought to have enhanced fishing as well as hunting, gathering, and gardening. Middle Woodland people developed more sophisticated fishing techniques with the use of gill nets, harpoons, and fishing hooks and lines. For several tribes, the waterways thus became increasingly important as sources of food and as travel and trade routes.

Likewise, Wisconsin's waters have since been vitally important to all the people who have occupied and developed the state's land. Water will always be essential to healthy living and healthy economies in Wisconsin. Wise use and management of the state's water resources will be a critical part of creating a sustainable future for coming generations and for all the other forms of life that share the water resources with us. Toward that end, it might be helpful for us to understand how Wisconsin's amazing waterways came to be what they are today, which is the focus of the following chapters.

2.1 Northwestern Wisconsin. MAPPING SPECIALISTS, LTD., FITCHBURG, WI

Lake Superior

DOUGLAS

Bois Brule River

BAYFIELD

IRON

ASHLAND

VILAS

St. Croix River

BURNETT

WASHBURN

SAWYER

ONEIDA

PRICE

POLK

BARRON

RUSK

LINCOLN

Upper Wisconsin River

TAYLOR

CHIPPEWA

Chippewa River

ST. CROIX

DUNN

MARATHON

PIERCE

PEPIN

EAU CLAIRE

CLARK

PORTAGE

WOOD

BUFFALO

N

JACKSON

0   30 mi.

0   30 km

TREMPEALEAU

MONROE

JUNEAU

ADAMS

2

THE NORTHERN HIGHLANDS

The face of the Northern Highlands—the region that includes much of northwestern and north central Wisconsin—is mostly a product of the glaciers, but the roots of its story go back to Precambrian time. The landscape was shaped first by the Midcontinent Rift, which formed a wide swath of hard basalt bedrock in the northwest corner of the state. Next, the formation of the Wisconsin Dome determined the directions of streams and rivers in the north, which flow generally southwest, south, or southeast from the center. Ancient inland seas deposited layers of sandstone, shale, and dolomite over the entire area, and since then, erosion has scoured the topmost layers off the dome, leaving Cambrian sandstone bedrock across most of the highlands.

Still, it was the glaciers that shaped today's Northern Highlands. These masses of ice and their voluminous meltwaters completed the job of removing dolomitic rock layers and left an array of moraines, drumlins, kames, tunnel channels, and eskers overlying the Cambrian sandstone bedrock. As they melted back, the glaciers calved icebergs and dropped a massive blanket of till to create vast tracts of hummocky land peppered by stranded chunks of ice, some of them miles in diameter, that melted out over centuries to form kettles, which later became hollows, wetlands, and lakes.

Those lakes—which make up the lion's share of Wisconsin's 15,000-plus lakes—are a key feature of the Northern Highlands. Geologist and author Robert Thorson called them the "signature landforms" of the upper Midwest, akin to the red rock canyons of the American Southwest and the volcanic mountains of the Pacific Northwest.[1]

Another legacy of the glaciers is a steady, gentle uplifting of land in the far north, known as isostatic rebound. Earth's crust—which floats on a quasi-liquid

2.2 The Lake Superior shore at Big Bay State Park on Madeline Island

mantle, much like barges on a river—was pushed down in this area, like a fully loaded barge, by the dense, heavy glaciers. Since the retreat of the glaciers, that land has rebounded, in some areas by tens of feet, and is still doing so today. This has made for some oddly configured glacial features, such as ancient beach lines that gradually slant upward as they run to the north.

In this chapter, we explore sites that feature many of these phenomena, starting with the south shore of North America's largest lake, followed by four key rivers and a major share of Wisconsin's lake country, all part of the glacial legacy (Figure 2.1).

THE LAKE SUPERIOR SHORE

Lake Superior, the largest freshwater lake in the world when measured by surface area, contains more water than the other four Great Lakes combined. Its surface spans 31,700 square miles, and it collects runoff from a land area of

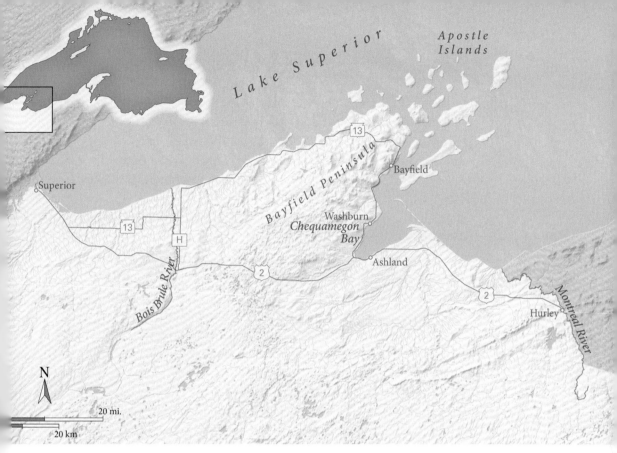

2.3 Wisconsin's Lake Superior shore. MAPPING SPECIALISTS, LTD., FITCHBURG, WI

more than 49,000 square miles via more than 300 streams and rivers that feed the lake. Its shoreline runs a rugged 1,729 miles, and its island shorelines total 997 miles. Wisconsin's share of the lake's shorelines includes about 200 miles of the lake's south shore and 200 miles of the Apostle Islands' shorelines (Figure 2.2). Thousands of people travel to Wisconsin's Lake Superior shore every year to enjoy its picturesque beaches, towns, apple orchards, islands, and waterfalls (Figure 2.3).

In its current form, Lake Superior is a youngster—only a few thousand years old. The lake sits on the south side of the Canadian Shield, a vast plate of rock centered on Hudson Bay in eastern Canada. The shield was formed beginning around 2.6 billion years ago by magma rising from the Earth's mantle into its crust. The magma cooled to form granite and related types of rock, which became the basement rock—the base on which all other rock layers were later deposited—and the core of the North American continent.

While that core has remained largely unchanged over time, a geologic

drama in several acts has taken place on its stage. Tectonic forces led to a series
of continental collisions that added to the size of the Canadian Shield. Such
forces stretched the crust in some areas, warped it in others, and fractured it in
still other locations. Mountain ranges rose, possibly to great heights, and over
eons, erosion leveled them into a rocky, rolling plain with low-profile basins
and highlands.

One such basin, called the Animikie Basin, was invaded by an ancient sea
around 2.2 billion years ago. That sea spread west from what is now northern
Minnesota across northern Wisconsin and the future site of Lake Superior and on
across northern Michigan to southeastern Ontario. The shallow waters around
the shores of the Animikie Sea deposited iron-rich sediments for some 200 million years.[2] This apparent one-time deposit of iron had to do with increasing
levels of oxygen in the atmosphere, which changed the ocean chemistry. This
made it possible for dissolved iron to settle out of ocean waters and to be stored
in sediments, which eventually became part of the body of rock, on the shores of
that ancient sea. These rich iron deposits played a major role in the development
of human cultures and economies of the area, especially in recent history.

Another major act in the drama of Lake Superior geology was the Midcontinent Rift and the formation of the Lake Superior syncline, covered in detail
in Chapter 1, around 1.1 billion years ago. The syncline is defined by the cliffs
along its north side in Minnesota and Ontario and, on its south side, by the highlands of Wisconsin and Upper Michigan. From both the Northern and Southern
Highlands, the basement rock slopes down thousands of feet, bottoming out on
the central axis of the lake bed.

The end of the Keweenawan era of rifting and volcanic activity, somewhere
around 1.075 billion years ago, brought a period of relative stability to the region, but the rift episode left its mark. Within the huge swath of newly formed
basalt were deposits of minerals that would be valuable to future human residents of the Lake Superior region. Copper, silver, gold, and other minerals rose
with the magma from deep in the earth and were deposited within crevices and
openings between layers of ancient rock underlying the younger basalt layers.

The Keweenawan episode also produced the abundance of Lake Superior's
famous agates—those colorful, semiprecious stones now hunted by rock hounds
on the shores of the lake. As the basaltic lava erupted and cooled on the surface, gas bubbles rose and formed tubes and cavities within the hardening rock,
which gradually became coated with layers of minerals varying in shades of tan,

2.4 Sandstone deposited 500 to 600 million years ago decorates the shores of the Apostle Islands.

brown, red, and white. These layered bodies of minerals hardened and eventually were exposed and broken apart by weathering and erosion. Along the lake's shores, waves have since tumbled these broken fragments against one another and among other rocks, naturally polishing them over the centuries to form what we now call agates.

During the long period following the Keweenawan eruptions, the mountains that were formed by continental collisions in early Precambrian time continued to erode. Grain by grain, they were pulled down by wind and flowing water. Several lakes formed in the nascent Lake Superior basin, and streams brought eroded sand and gravel to the ancient shores. Over many millions of years, they built thick layers of sandstone and conglomerate, including the red sandstone that famously lines the Bayfield Peninsula and the Apostle Islands (Figure 2.4). This Lake Superior Sandstone is stained reddish brown due to the presence of iron deposited by the Animikie Sea.

The quiet period of building by deposition ended about 900 million years ago when primitive North America collided with another continent at its southeastern edge, hundreds of miles from the rift zone. The compression of the crust caused blocks of land (fault blocks) to be thrust upward, creating landforms called horsts. One such fault block, called the St. Croix Horst, was centered on the St. Croix River and bounded on the northwest by the Douglas Fault—the line along which the block of land was heaved up above the area northwest of it.

The horst continued to rise for around 10 million years, and geologists estimate that on some stretches of the Douglas Fault, it rose many thousands of feet, lifting the layers of basalt buried deep beneath the Lake Superior Sandstone above the younger sandstone layers to the northwest. The fault heaved up slowly—a few inches per century—and erosion of the rising ridge kept it from getting very high above the surrounding plain. Today, the Douglas Fault is mostly level with the land around it, except in a few places where it forms prominent cliffs. A number of streams flow over such escarpments, creating scenic waterfalls.

The next major act in Lake Superior's geologic drama was the coming of the glaciers, beginning about 2.5 million years ago. By that time, the future lake bed was a vast, shallow basin drained by a river flowing northeast on a forested sandstone plain. Several times, the ice sheet crept down from the north, and the ancient river valley channeled the ice along a northeast-to-southwest route. Each passage of the massive ice bodies crushed some of the softer sandstone underlying the valley, picking up the resulting stone fragments and loads of sand and moving them along. Each successive advance of ice gouged out more Precambrian sandstone and made the future lake basin deeper. At the southwest end of the lake basin, however, the glaciers slowed, thinned, and lost erosive power, leaving a large mass of sandstone as the foundation for the Bayfield Peninsula and the Apostle Islands.

This sandstone mountain stood in the path of the most recent ice mass—a lobe of the Wisconsin glaciation. It divided that lobe into the Superior and Chippewa Lobes, which moved along the mountain's northwestern and southeastern flanks, respectively, while chewing away at its northeast end, as previous glaciers had done. The massive erosion created the Apostle Islands, which had been part of the ancient Bayfield Peninsula. The actions of both the glacial ice and the fast-flowing meltwater under and around the melting glacier carved out some of the low-lying areas that now separate the islands from the peninsula and from one another (Figure 2.3).[3]

As the glacier retreated, its aftereffects continued to change the shape of

Wisconsin's Lake Superior shore. The glacier left a moraine lying within 20 miles of the shore across northern Douglas, Bayfield, Ashland, and Iron Counties. A huge amount of that glacial till now sits on top of the high ridge of Bayfield Peninsula along with an enormous mass of sandy outwash, the material deposited by broad streams flowing away from the melting glacier. The outwash on the peninsula is up to 500 feet thick in some areas.[4]

Another effect of the retreating glacier was the formation of glacial lakes—bodies of meltwater in low-lying areas dammed on one side by the retreating wall of ice. Such lakes formed on either side of the Bayfield Peninsula and eventually merged into one large body called Glacial Lake Duluth, an ancestor of Lake Superior. Over the following centuries, the lake level rose and fell, getting as high as 500 feet above Lake Superior's present level. At its highest levels, the lake's icy waters lapped against the moraine that lay inland from the present-day south shore of the lake. That area between today's shore and the moraine is called the Lake Superior Lowland, and there the glacial lake dropped the thick layers of heavy reddish clay, silt, and sand that now make up much of the lake's south shore. Those sediments are reddish because they are made from the Precambrian sandstone that was stained red by iron oxides 2 billion years ago and then pulverized and hauled to the shore by the glaciers.

When the last glacier had finally withered and the big lake slowly dropped to its current level, the muddy red Lake Superior Lowland lay glistening in the sun, stretching away from both sides of the freshly carved sandstone peninsula and islands. The Wisconsin shore we know today was slowly beginning to take shape.

At Lake Superior's west end, called the lakehead, the shape of the shore was affected by a phenomenon known as differential rebound. The land had been compressed by the glacier and began to rise back up, but the rate of rebound was slightly higher at the lake's east end, so the entire lake basin began gently tilting to the west. (This process is still occurring; the east end of the lake is rising roughly a half foot faster per century than the west end.) Thus the water on the lake's west end has risen slowly while the water at the east end has dropped. As a result, what was once the valley of the St. Louis River flowing into the west end of the lake between present-day Duluth and Superior is now St. Louis Bay, the mouth of the river pushed underwater by differential rebound.[5]

Another feature that evolved fairly quickly was a lakehead sandspit stretching between Duluth and Superior. It got its start about 3,000 years ago when large waves sweeping into shallow lakeshore waters scooped sand from the

2.5 On its way to Lake Superior, the Black River crashes over the Douglas Fault in Pattison State Park's Big Manitou Falls.

bottom and built a sandbar. The sandbar grew until it emerged from the water, and then wind and near-shore water currents helped to grow and shape it. It is now a nearly continuous sandspit across the end of the bay, except for gaps near the Wisconsin end and at the ship canal in the Duluth Harbor. Now known as Park Point on the Minnesota side and Wisconsin Point on the other side, it is nearly 10 miles long, up to a quarter mile wide, and 10 to 20 feet above the lake level, hosting an airport and many homes.

Along Wisconsin's Lake Superior shore, beginning about 9,500 years ago, postglacial rivers draining the highlands to the south quickly cut through the glacial lake sediments and glacial till and found bedrock. As they flowed north into Lake Superior, some of them crashed over the basalt escarpment of the Douglas Fault. While they have been wearing away at that rock for millennia, the hard rock has resisted and today still forms spectacular waterfalls such as Big Manitou Falls in Pattison State Park (Figure 2.5).

At the center of Wisconsin's shore, the retreating glacier left Bayfield Peninsula jutting northeast along the axis of the lake basin. Composed of late Precambrian and early Cambrian Lake Superior sandstone, the peninsula is flanked by 22 islands made of the same type of stone. The hilly peninsula and archipelago display the beautiful sandstone that ranges in color from red to tan to brown. Steady erosion by lake waves has kept the stone vibrantly colored while carving caves, pillars, and arches along the shorelines, making the peninsula and islands highly attractive to travelers throughout the centuries since glacial times (Figure 2.4).

The southeast side of the peninsula, being more sheltered from westerly winds, lacks the wave-carved features of the northwest side, but instead is partly lined by sandy beaches. This side of the peninsula is defined by the deeply indented Chequamegon Bay, which measures 12 miles from head to mouth (Figure 2.3). The head of the bay sports a broad sandy beach that figured prominently as a historic settlement and meeting place among Native peoples and European explorers and fur traders. Today, it forms a popular swimming beach for the city of Ashland.

Owing to the formation of a sandspit stretching most of the way across its mouth, Chequamegon Bay functions as a natural harbor. In postglacial days, the bay was larger than it is now, with water covering a stretch of the shore from just west of Ashland to the mouth of the Bad River, a swath of land up to three miles wide. The site of Ashland was underwater. Big lake waves hitting the shallow mouth of that bay moved sand from the bottom to form a string of sandbars that grew higher and longer due to wave and lake current action. Eventually, the sandbars connected and formed a spit, now called Chequamegon Point. Long Island, which lies off the point, was once part of the spit, but Lake Superior's ferocious storms created the gap that now separates Chequamegon Point and Long Island.[6] Over centuries, the calm, shallow waters southeast of the spit were filled with sediments by inflowing rivers and colonized by plants to become marshland and sloughs as Chequamegon Bay was narrowed.

From today's Chequamegon Bay east to the Michigan border is another stretch of Lake Superior Lowland, starting with the generally flat area that was once part of the bay. This coastal area slopes gently up from shoreline cliffs of red clay sitting 10 to 20 feet over the water to highlands up to 500 feet above the lake level. Like the lowland west of the peninsula, this one is dissected by several streams that, since glacial times, have steadily carved shallow valleys as they flow to the lake. Some streams have worked their way down to Keweenawan sandstone and basalt bedrock, where they flow over ledges and cliffs, forming arresting rapids and waterfalls.

Unlike inland Wisconsin, the shore area was, and still is, affected by the narrow range of water temperatures in the big lake. The massive body of water takes so long to warm up or cool down that the state's extreme seasonal air temperatures are moderated at the shore. As the climate warmed after the most recent glacier shrank away, Wisconsin's Lake Superior shore area went through the succession of plant communities typical of postglacial landscapes—from tundra to shrubland to black spruce forests and later to a mix of pine and hardwood forests. Before the large-scale logging of the 1800s, forests on the south shore and islands were dominated by hemlock, white pine, sugar maple, yellow birch, and white birch. Balsam fir and white cedar took hold in low, wet forest areas and on rocky, windy shorelines.[7]

With these postglacial changes came a succession of animal communities starting with animals that preferred the tundra, including caribou, musk oxen, and wooly mammoths. Hunters of the Paleo-Indian tradition who favored big game arrived in the Lake Superior area around 9,000 years ago. The forests eventually hosted diverse animal communities that included white-tailed deer, wolves, and smaller mammals.

Over the centuries, Archaic and Woodland peoples hunted in the forests flanking the Lake Superior shore, fished along the shore and its tributary rivers, and gathered plants, berries, and nuts from the forests. They did some gardening, but due to the short growing season in northern Wisconsin, they relied more on gathering and fishing during summer and on hunting year-round for their food sources. Between 7,000 and 3,000 years ago, the Archaic people developed tools, including the ax, adze, gouge, and fishing hooks and nets, which allowed them to form semipermanent settlements, and they widened their trade networks with other tribes. Late in this time span, they also developed the bow and arrow, which allowed for much more efficient hunting.

The Copper culture appeared at this time, overlapping the Archaic and Woodland cultures. The Woodland people lived in the area between 3,000 and 1,000 years ago. One key advance of the time was the development of pottery, which allowed for food storage, more varied diets, and increased trading. Beginning about 1,200 years ago, the people living on Wisconsin's Lake Superior shore began harvesting wild rice, which became a lasting staple in their diet and remains so today.

The most prominent tribe living along the Wisconsin shore when Europeans showed up in the 1600s was the Ojibwe, or Anishinaabe (also called the Chippewa). Their name for Lake Superior was *Gichigami*, meaning "Great Lake" (from which comes the more commonly reported variation *Gitchee Gumee*). The lake's common English name was derived from Lac Superieur, by which the early French explorers referred to this uppermost of the chain of Great Lakes.

Prior to the 1600s, the Ojibwe had moved into the area from the east and were living as seminomadic bands in seasonal camps, their way of life being one of subsistence. Their population along the shores of Lake Superior when Europeans began arriving is estimated to have been around 100,000.[8] In the centuries following European exploration and settlement, the Ojibwe were gradually forced off the lands they occupied around Lake Superior. By 1854, the US government had reduced the Ojibwe lands to eight reservations in the Lake Superior region. Five of them are located on the lake's shores, amounting to a small fraction of the shore land they and their ancestors had occupied for centuries.

TRAVEL GUIDE
The Lake Superior Shore

This travel guide generally follows Highways 2 and 13 along the Wisconsin shore of Lake Superior. Coming in from Minnesota on US Highway 2, you cross over St. Louis Bay on the Richard I. Bong Bridge, entering the zone of the mended rift that once nearly split the continent in two. The rift is now buried under many thousands of feet of basalt and sandstone lying beneath the waters of the bay. Overlooking the bridge and bay from the Minnesota side is the northwest wall of the rift zone—the basalt cliffs of Duluth and the famously scenic palisades of Lake Superior's north shore stretching away from Duluth. Wisconsin's counterpart to Minnesota's cliffs and palisades is the escarpment just south of the Lake Superior lowland, formed partly by the Douglas Fault and partly by a glacial moraine, both on display in Pattison State Park.

In the park, about 10 miles south of Superior on State Highway 35, you can take short hikes to view two waterfalls—Big Manitou (Figure 2.5) and Little Manitou Falls. At both sites, the Black River plunges over ledges of 1.1-billion-year-old basalt that were heaved up along the Douglas Fault millions of years ago. Big Manitou Falls, with its 160-foot drop, is Wisconsin's highest waterfall, the fourth highest east of the Mississippi, and the most dramatic display of the Douglas Fault. The Beaver Slide Trail follows the river upstream to Little Manitou Falls—a shorter drop but also spectacular. The trail crosses a body of sandy soil and gravel—an ancient beach where Glacial Lake Duluth sat for centuries as it drained after the glacier retreated.

Highway 2 runs along the lakeshore through the city of Superior and then departs from the shore for a few miles. In this flat area, the highway runs over reddish soil made mostly of clay and sand laid down by the glacier and by Glacial Lake Duluth. The highway curves from southeast to east and parallels the lakeshore for several miles.

Just after it curves east, Highway 2 is intersected by County Highway U, which leads you north a short distance to Amnicon Falls State Park. Within the park's borders, the Amnicon River, draining a large bog that lies to the south, drops 180 feet on its way to the Lake Superior shore. Three of the park's four waterfalls flow over the escarpment of the Douglas Fault. At the Upper Falls (Chapter 1, Figure 1.1), water flows over the ancient basalt ledge lifted by the fault over the much younger Lake Superior Sandstone downstream. Nearby on another branch of the river, Snake Pit Falls also plunges over the Douglas Fault escarpment. (When the river is high, this is a dangerous waterfall; mind children and pets closely.) Lower Falls drops over a resistant ledge of Lake Superior Sandstone laid down in late Precambrian and early Cambrian time. Viewers of the falls can easily see how this sandstone is bedded in flat layers, each deposited over many thousands of years by ancient seas that covered the area.

Back on Highway 2, continue for several miles as the road crosses the Lake Superior Lowland and the ancient lakeshore of Glacial Lake Duluth, rising above and dipping below the highest level of that lake several times. If the clock were turned back suddenly to the time just after the glacier receded, an amphibious vehicle would come in handy, for this path would lead into and out of the lake several times. The road rests on red clay laid down by the lake waves that rolled over the area all those thousands of years ago.

At the town of Brule, 24 miles east of Superior, the Bois Brule River (discussed later in this chapter) flows from the south toward the big lake. From Brule, take County Highway H north from Highway 2 for eight miles. There it intersects with State Highway 13, which takes you north to the Lake Superior shore and then runs east along the shore and around the Bayfield Peninsula. The first town on the peninsula is the old fishing village of Port Wing. On its west side is the turnoff to Twin Falls Park, where Larson Creek flows over outcroppings of reddish sandstone in a pair of waterfalls, dropping into a picturesque glen. Over millennia, the creek has carved bowls out of the mass of sandstone where hikers can view cross-section exposures of the layered stone, some of which have been fractured and tilted by centuries of frost action and spring flood erosion.

Beyond Port Wing, take Highway 13, which angles northeast and ascends and wraps around the flank of one of the many sandstone hills that were carved, scoured, and rounded off by glacial ice and meltwaters flowing across the peninsula. As the road descends again, six miles up the shore from Port Wing, it runs through the small town of Herbster in the valley of the Cranberry River. There, Bark Point Road runs northeast from Highway 13 to the end of Bark Point, a peninsula jutting into the lake. About halfway to the point, you can take Bark Bay Road east from Bark Point Road. It passes Bark Bay Slough, a state natural area along the southwest side of Bark Bay with a picturesque lagoon defined by a sandspit at the head of the bay that hosts a fragile plant community crowned by a row of conifers. This is where bog and fen ecosystems have gradually filled part of the shallow bay, taking over what once was a rocky lakeshore. From a parking area on the lagoon, on a windy day, you can hear the waves crashing on the other side of the sandspit, even though the lagoon remains calm. From here, follow Bark Bay Road easterly to meet Highway 13 about eight miles east of Herbster.

A little farther east, Highway 13 descends into the valley of the Siskiwit River where Wisconsin's northernmost town, Cornucopia, overlooks Siskiwit

2.6 Lost Creek Falls, a few miles upstream from where the creek flows into Lake Superior. GAIL MARTINELLI

Bay. This shallow and well-protected bay is popular among beachgoers and boaters, and the town serves as a port for commercial fishing vessels. Just west of town is the Lost Creek Bog State Natural Area, a sensitive and increasingly rare lakeshore estuarine ecosystem where rising lake levels have put the mouths of three creeks underwater—evidence of the slow rise of the waters along the western Lake Superior shore due to differential rebound of the land since the glacier departed. These three creeks descend from the hills to the south, and one forms a gorgeous, secluded waterfall called Lost Creek Falls (Figure 2.6). A side trip there involves a 1.5-mile hike—highly recommended. Take County Road C up the hill out of Cornucopia for a little more than two miles to the trailhead parking area, a short distance west of the highway.

Out of Cornucopia, Highway 13 turns east across the peninsula, climbing steeply into the upper reaches of its hills. The highway departs the shore at the west end of a 12-mile stretch of the Apostle Islands National Lakeshore, a federally protected length of the shore famous for its intricately carved caves, pillars,

2.7 Ice floes during winter create the Bayfield Peninsula's famous ice caves. DAVE OLSON, WASHBURN, WI

and arches in colorful Lake Superior Sandstone. Some of this stretch displays stunning floes of ice during winter, making the Bayfield Peninsula Ice Caves a world-renowned destination for hardy winter hikers and kayakers (Figure 2.7). Not far from Cornucopia, Myers Beach Road goes to a popular departure point for lake kayakers and lakeshore hikers. From there, a rugged two-mile trail runs to the cliffs that form the famous ice caves.

At the east end of the National Lakeshore is the western border of Red Cliff Indian Reservation, the remaining land of the Red Cliff Band of Lake Superior Ojibwe. It includes 60 square miles of the shoreline watershed, 46 miles of streams and rivers, and protected wetland sloughs that for centuries have been an important source of food and other resources for Indigenous people. The reservation wraps around the northeast end of the peninsula. Highway 13 mostly skirts the reservation but enters it west of the town of Red Cliff where the road veers south along the lakeshore. Here the Red Cliff Band operates a casino and other tourist attractions, including its annual July Fourth Powwow and Celebration.

Just south of Red Cliff is the city of Bayfield, a popular destination for vacationers, known for its busy marina, historic Victorian mansions that provide upscale lodging, and the Apostle Islands National Lakeshore Visitors Center. The National Lakeshore and Apostle Islands have the same degree of federal protection as do other national parks. Bayfield serves as the departure point for visitors to the islands, including the largest, Madeline Island, which is developed and inhabited year round but is not part of the National Lakeshore. A ferry service provides regular daily transportation to Madeline Island.

If you want a colorfully historic, naturally beautiful, and downright quirky place to visit, Madeline Island is a good destination. It is a non-wilderness experience, unlike the other islands, but its village of La Pointe, where the ferry boats dock, is like no other town, having a unique variety of restaurants, inns, bars, and shops. The Wisconsin Historical Society runs the Madeline Island Museum, a rich collection of artifacts and information about the deeply rooted Ojibwe culture and the European fur trade. Also on this island is Big Bay State Park, where you can experience a wide variety of ecosystems, including a sweeping pristine beach (Figure 2.2), a rare and fragile lakeshore bog and lagoon (see photo on page x), and spectacular displays of 600-million-year-old red sandstone lying along the shore under an ancient cedar forest (Figure 2.4).

For the wilderness experience, you can catch limited ferry service to some of the other islands for hiking and/or camping. Experienced lake paddlers travel among the islands in sea kayaks. During the late 1800s and early 1900s, loggers cleared all but five of the islands, and three islands were quarried for sandstone (popularly known then as brownstone and used to construct hundreds of buildings in several cities). Most of the islands were sparsely populated.

In spite of all that historical human activity on the islands, each one except Madeline has largely been reclaimed by natural forests and wetlands. Because they lie in the tension zone between boreal and temperate forests, the islands host rare plant assemblages, including clay bluff, lagoon, bog, and dune communities. Stands of old-growth mixed conifer and hardwood forests also have survived on a few of the islands. Animal communities on the islands are similarly diverse, but most islands do not host white-tailed deer. Consequently, the ground cover of native plants on these islands has remained undisturbed, reflecting the island ecosystems that existed before Europeans arrived.

South of Bayfield, continue on Highway 13 as it flanks the peninsula's high sandstone hills and hugs the lakeshore. Mount Ashwabay, site of the popular Big Top Chautauqua and other attractions, sits more than 500 feet above the lake, topped by an enormous mass of sandy outwash from the retreating glacier. The next city down the shore is Washburn, once the center of a thriving quarrying industry, now a center for the arts and historic preservation situated on the western shore of Chequamegon Bay. From Washburn, the road follows the shore for another four miles, where it meets Highway 2. To continue the lakeshore trip, turn east onto 2. Less than half a mile west of this junction on Highway 2 is the Northern Great Lakes Visitor Center, which houses a superb collection of artifacts and information about the geology, ecology, history, and cultures of the region. East of the junction, Highway 2 follows the lakeshore two miles to the city of Ashland.

Ashland was once a major center for shipping on the shore. During the first half of the 20th century, busy railroad lines carried iron ore from the mines of the Gogebic Range, southeast of the city, to the Port of Ashland. Trains hauled the ore onto four huge docks where ore boats took on their loads for shipping on the St. Lawrence Seaway. With declining shipping of iron ore, the docks fell into disuse, and in 2016 the last of them was dismantled down to its sprawling foundation, which is now a city park. Ashland is a center for health care, and its Northland College, established in 1892, has become a world-class institution of environmental education.

East of Ashland, Highway 2 continues along the shore. Not far from the city, the road traverses the Bad River Indian Reservation for 30 miles. This is where the Bad River Band of Lake Superior Ojibwe were forced to move from surrounding lands they had occupied for centuries as a result of a treaty in 1854. The reservation contains the Kakagon and Bad River Sloughs, two of the richest estuarine ecosystems in the world. The river delta containing the sloughs is a center for biodiversity hosting wild rice beds, fish spawning grounds, and water-fowl habitats that have provided resources to the Ojibwe for thousands of years.

Leaving the reservation, the highway climbs a long, straight route up steep Birch Hill. At the top of the hill, the area north of the road is a long, narrow highland that once protruded into Glacial Lake Duluth as a sandy peninsula. There, geologists have found evidence of high sand dunes that once stood on that ancient lakeshore and are now covered completely by forest.

From the top of Birch Hill, Highway 2 runs two miles to the junction with State Highway 169, which provides an interesting side trip to Potato River Falls. Going south a few miles to the old town of Gurney, Highway 169 lies on the flat, sandy surface of an old delta—the place where the ancient Potato River flowed into Glacial Lake Duluth when its waters covered the land to the north. On the south side of Gurney, Falls Road heads west to the site of the falls, a spectacular 90-foot cascade over ledges of conglomerate, shale, and sandstone deposited between a billion and 600 million years ago. From the parking lot, trails to the head (Figure 2.8) and foot of the falls provide splendid views.[9] If you wish to see more impressive waterfalls, continue south on Highway 169 for 15 miles to Copper Falls State Park (see frontispiece photo just before the title page).

After its intersection with Highway 169, Highway 2 continues east for five miles before its junction with State Highway 122 and the village of Saxon. To the north are Saxon Harbor, Wisconsin's easternmost lake harbor, and access to two gorgeous waterfalls on the Montreal River, which here forms a sinuous border between Wisconsin and Upper Michigan. The river tumbles over red Lake Superior Sandstone laid down upon 1.1-billion-year-old basalt after the Midcontinent Rift was halted. This area is sacred to the Ojibwe, who call the river *Kawasiji-wangsapi*, or "white falls river."[10] To reach the first waterfall, Saxon Falls, go north on Highway 122 for two miles and turn east on County Road B, traveling for nearly two miles to where the road veers south, and there turn north on the dirt road that leads to the falls. From the parking area, it is a half-mile hike to the falls.

2.8 Upper Potato River Falls

Access to the second waterfall, Superior Falls, is a few miles farther up Highway 122. Cross over the Montreal River into Michigan and go another half mile to a gravel road on the left leading to a parking area. From there, take a short walk to the edge of the lakeshore bluff, which is at least 100 feet high and made of red clay, sand, and gravel deposited by the glacier and by Glacial Lake Duluth. To get to the falls, take the steep path down to the lakeshore and the mouth of the Montreal River (Figure 2.9). Hike upstream within a deep gorge which itself is a sight to behold. Jumbled layers of ancient bedrock are displayed in its high, massive walls. A few hundred yards downstream of the falls, you will see them. The river drops 65 feet as it gushes into the gorge. Superior Falls lies at the eastern end of Wisconsin's Lake Superior shore.

2.9 The mouth of the Montreal River on the Lake Superior shore. Just upstream, the river clambers over Superior Falls.

THE ST. CROIX RIVER

If you could fly straight south from the Lake Superior shore, a few miles east of the city of Superior, you would see the land rising gradually but steadily for 10 miles. There, you would begin crossing a 20-mile-wide swath of forested hills before seeing the land below flatten into an area of wetlands and lakes. This level area forms a divide on which waters gather in rivulets and streams that flow either north toward Lake Superior or southwest toward the Mississippi

2.10 The Dalles of the St. Croix River

River valley. Here in this broad, lush wetland area lie the headwaters for two of Wisconsin's best-known rivers—the north-flowing Bois Brule and the southwest-flowing St. Croix River (Figures 2.1 and 2.10).

The St. Croix River drains a gently varied landscape and drops just 400 feet in its 154-mile journey to the Mississippi. In the placid peat bogs of its headwaters, tannic acid stains the water reddish-brown. The river flows languidly for most of its 25 miles within Wisconsin borders and the remaining 129 miles along the Wisconsin–Minnesota border.[11] However, the placid nature of the river belies a tumultuous geologic history that formed the St. Croix Valley.

That story begins 1.1 billion years ago with the formation of the Midcontinent Rift. Under most of the Upper St. Croix Valley (the stretch upstream of State Highway 70), the bedrock slopes northwest into the Lake Superior Syncline.

However, overlying the dipping bedrock are thick sediments from centuries of glaciation, which slope southwest, causing the St. Croix River to follow this slope from its headwaters.[12]

Between 1 billion and 900 million years ago, after the rifting period had ended, the primitive North American continent collided with another continent hundreds of miles southeast of the rift zone. This collision compressed the crust and caused a broad, narrow block of land to be thrust upward, squeezed between the masses of crust on either side of it. Such a formation is called a horst, and this one—centered on the St. Croix River—is known as the St. Croix Horst.

Along with the creation of the horst and associated shifting of land along faults, millions of years of erosion made a complex terrain of the layered basalt bedrock. Along one stretch of today's St. Croix River, centered around the city of St. Croix Falls, these forces cut gorges and shaped hills and cliffs that were up to 300 feet high. In the Cambrian seas around 500 million years ago, the rocky hills were islands and the gorges were lagoons bounded by the high cliffs. These seas and the rivers that flowed to them deposited layers of sandstone in the low areas among the islands.

The region was tropical at the time, lying 10 degrees south of the equator,[13] and is thought to have been wracked by occasional storms. Geologists draw this conclusion partly on the basis of the presence in the St. Croix Valley of conglomerate made of basalt boulders and stones cemented within sandstone. These stones and boulders likely fell from basalt cliffs lashed by storm waves. The boulders plunged into the Cambrian sea surf where they were tumbled and rounded by waves and currents for centuries before being buried in sand and then incorporated into conglomerate.

Ancient seas advanced and retreated in the area several times over hundreds of millions of years. When the last of them finally departed Wisconsin, possibly around 260 million years ago, the region was mantled by thick layers of sandstone, conglomerate, and dolomite. Erosion then took over as the dominant geological force and, in northwestern Wisconsin, eventually exposed deep layers of Cambrian sandstone and much older basalt. By the time Pleistocene glaciers began inching down out of Canada, the St. Croix River Valley region was mostly a rolling, forested plain on sandstone.

Of the six lobes that covered Wisconsin during that last glaciation, it was the Superior Lobe (Chapter 1, Figure 1.4) that most affected the St. Croix River Valley. Between 13,000 and 14,000 years ago, the Superior Lobe flowed generally southeast and covered northwestern Wisconsin in hundreds to thousands of feet of ice. It deposited vast quantities of sand, gravel, and boulders, which partially filled the ancient St. Croix Valley. Then, about 12,000 years ago, not long after the Superior Lobe had retreated from the area, a smaller lobe of ice invaded the valley from the west. It was a sublobe of the Des Moines Lobe (which covered parts of Minnesota and Iowa) called the Grantsburg Sublobe, named for the Wisconsin town located on the line of its farthest advance. This sublobe moved northeast out of Minnesota and crossed a segment of the St. Croix River Valley between today's Highway 70 bridge and the area just downstream of Osceola.[14]

Geologists use a number of clues to determine how and where ice masses once flowed. For example, they can identify different types of till, the fine debris carried and dropped by glaciers as they melted. The Superior Lobe left reddish, sandy till, which had been scraped from the basin of Lake Superior and carried south. The Grantsburg Sublobe carried a gray, silty till from the lowlands in Minnesota. In the St. Croix River Valley, that gray till overlies the reddish till, showing that the Grantsburg Sublobe arrived after the Superior Lobe departed.

Glacial lakes, formed by water from the melting glacier, played a huge role in the formation of today's St. Croix Valley. First, as the Superior Lobe retreated northward beginning 13,000 years ago, Glacial Lake Lind formed in an oblong lowland stretching from the area of Danbury, Wisconsin, southwest toward Minneapolis, centered on the town of Lind, Minnesota.[15] This is thought to be the valley of an ancestral version of the St. Croix River. The icy waters of Lake Lind reached depths of 180 feet and sat in place for more than 1,000 years.

As the Superior Lobe retreated, great volumes of sand and gravel washed into Lake Lind in a river, building a delta that replaced the lake and buried much of the ancient river valley. Overlying this outwash today is another type of sediment deposited by another glacial lake that formed against the retreating Grantsburg Sublobe. By about 12,000 years ago, the sublobe's ice had dammed the postglacial St. Croix River near the location of present-day Osceola, Wisconsin, forming Glacial Lake Grantsburg, which stretched from Osceola northeast into and beyond the center of Burnett County. It lay over the sediments left by the postglacial river delta that had buried Lake Lind. Lake Grantsburg sat for about a century before draining out of its southeast side, likely forming today's Lower St. Croix River Valley.

However, the greatest postglacial drama was yet to occur in the new St. Croix Valley. While Glacial Lakes Lind and Grantsburg had been filling the St. Croix

Valley with their sediments, the Superior Lobe had retreated into the Lake Superior basin, and its meltwaters formed Glacial Lake Duluth. That lake's waters rose to fill the southwest end of the Lake Superior basin, putting Wisconsin's present-day Lake Superior shoreline under 500 feet of water. Glacial Lake Duluth lapped against the highland not far from the flats that today hold the headwaters of the Bois Brule and St. Croix Rivers. In that headwaters area, the glacial lake's waters were held back by a fragile wall of ice and glacial debris.

When that dam was breached between 10,000 and 9,500 years ago, a gigantic torrent of icy water overwhelmed the headwaters area and roared down the valley of the St. Croix. That flood bored a channel that is wider than a mile in most places and more than 100 feet deep.[16] The work of the floodwaters is most dramatically on display at Interstate Park with its Dalles of the St. Croix, a canyon defined by sheer 100-foot-tall walls of basalt (Figure 2.10). Over a period of perhaps 500 years, two or more massive floods scoured off layers of sandstone overlying the basalt and tore into that harder rock. The floodwaters, channeled and concentrated by the basalt hills north of St. Croix Falls, pressed into the naturally occurring deep cracks in the basalt, chiseling and loosening large blocks and pushing them downstream, thus carving the narrow canyon. Strong eddies in the ferocious floods captured rocks and boulders and spun them against the bedrock for decades or centuries, drilling the potholes for which Interstate Park is known (Figure 2.11).

Geologists think this dramatic flooding took place more than once as the ice retreated and readvanced at least twice in the Lake Superior basin. In fact, the ancient St. Croix River may also have helped drain older glacial lakes during the 2.5-million-year Pleistocene epoch.

Over decades or possibly centuries of drainage, the valley had become partially filled with sediments deposited by the floodwaters. The river level dropped as the glacial lake drained to the north. In most places along today's St. Croix River, the water occupies just a small fraction of the width of the glacial floodwater channel. In the process of narrowing, the river has carved smaller channels within the ancient trench, leaving long, broad terraces along several stretches of the river. Modern river towns, including Taylor Falls, Minnesota, and the Wisconsin cities of St. Croix Falls, Osceola, and Hudson, are built on these terraces.

Centuries after the glacier departed, forests evolved and covered the river terraces and the adjacent hills of the St. Croix Valley. The Upper St. Croix River

2.11 A pothole (five feet across and about eight feet deep) drilled by postglacial flooding in Interstate State Park

formed a rough dividing line between two forest communities. To the northwest were forests of white and red pine growing in loamy soil. To the southeast were pine barrens, consisting of jack pine and prairie grasses in sandy soil. The Lower St. Croix River down to its confluence with the Mississippi hosted forests of white, black, red, and burr oak; sugar maple; basswood; elm; and in the lowlands, willows, soft maple, and ash.

Beginning with the retreat of the glacier, the river valley served as hunting grounds for the people who first lived in the area. Archeologists have uncovered bison bones along with arrowheads and early tools in Interstate Park, indicating that hunters roamed the area between 10,000 and 12,000 years ago. Over the centuries, Indigenous cultures evolved. Stories told by descendants of the early inhabitants of this area, supported by archaeological evidence, indicate that seminomadic peoples lived in seasonal camps within the St. Croix Valley. For centuries, these people were making maple sugar in the spring, fishing the river during summer, and harvesting wild rice from area lakes and wetlands in the fall. All the while, they hunted game in the forests. Later peoples learned to store food that would sustain them through the winters.[17] Groups living in the St. Croix Valley over the centuries include the Dakota (also called the Sioux), Ojibwe (also called Chippewa), Ho-Chunk (historically called the Winnebago), and Menominee.

As trade became a large part of the Native American way of life, the St. Croix River became an important part of the trade route between Lake Superior and the Mississippi River. It also served as a rough border separating the territories of the Dakota to the northwest and the Ojibwe to the southeast. These tribes developed a deep animosity toward each another, and each sought to drive the other out of the area. The St. Croix Valley was the scene of some major battles between them.[18]

The conflict between these nations may have been linked to the Ojibwe people's increasing involvement in the fur trade. In the early 1600s, the French and later the English realized the lucrative nature of trading goods such as blankets, kettles, and guns for help from the Ojibwe in gathering animal pelts, especially those of beaver. The St. Croix and Bois Brule Rivers became a major artery for the transport of furs between the Mississippi Valley and Lake Superior and points east in what was then New France. The fur trade was intensive—easily the largest commercial enterprise in the region for almost 100 years spanning the 17th and 18th centuries—and beaver populations in some areas were effectively exterminated.[19]

The St. Croix River has its name as a result of the fur trade—not geology, as was once thought. It was previously believed that the name came from a cross-shaped rock feature on one wall of the gorge at Interstate Park. Now, it seems clear from the records of 17th-century French explorers that the river, referred to as Rivière de Sainte-Croix, was named after a Frenchman named Sainte-Croix who traded with Indigenous people in the region.[20]

Following the fur trade, European immigration became the dominant force of change in the St. Croix Valley. European settlers backed by a growing military presence in the area crowded Native nations out of the valley and the adjacent flat uplands that attracted would-be immigrant farmers. After a long period of conflict, the US military forced the Ojibwe and Dakota to cede their lands in and around the valley in an 1837 treaty.

The St. Croix River and its major tributary, the Namekagon River, have long been recognized as natural treasures worth preserving. In 1968, at the urging of

Wisconsin Senator Gaylord Nelson, Congress passed the National Wild and Scenic Rivers Act. The Upper St. Croix and Namekagon Rivers were among the first rivers designated for special protection under this law, which led to strict limits being placed on the development of lands immediately adjoining the rivers.

TRAVEL GUIDE
The St. Croix River

Rather than describe a tour of the St. Croix River Valley from source to mouth, this travel guide highlights points of interest in the valley.

Headwaters and Solon Springs. Northeast of the town of Solon Spring off US Highway 53, at the junction of County Roads A and P, is the divide that separates the north-flowing Bois Brule River from the southeast-flowing St. Croix. For centuries, this was the site of a heavily traveled portage between the two rivers—an important link on a major trade route for Native Americans and later for European fur traders. (For more on the Brule–St. Croix Portage Trail, see the next section, "The Bois Brule River.") Nearby Solon Springs sits on a broad terrace made of outwash from the Superior Lobe of the glacier, consisting largely of fine, reddish sand that can be seen along many roads in this area, eroded from Precambrian sandstone to the north.

The Namekagon River. The Namekagon River flows into the St. Croix River just east of the community of Riverside on State Highway 35. On a map, the two rivers appear to be branches of one river. Indeed, that is how the early European river travelers regarded them, calling the Namekagon the East Branch of the St. Croix. The present-day name comes from the Ojibwe term *Namekaagong-ziibi*, meaning "river abundant with sturgeon."

Spending time on the Namekagon River makes for a wonderful side trip (Figure 2.12). The best way to see the river is to canoe it. (This river is near and dear to me—I canoed it many times while growing up in this part of the state.) The Namekagon flows across a vast plain of sand and gravel deposited by the glacier and its meltwaters. The river has washed away finer materials, revealing a gravelly and rocky bed. Watchful canoeists on the Namekagon may be treated to wildlife sightings, including white-tailed deer and black bear seeking a drink along the riverbank, river otters sliding into the water after fish or mollusks,

2.12 The Namekagon River is one of the original eight National Wild and Scenic Rivers managed by the National Park Service. PAUL F. OSTRUM

osprey and eagles soaring overhead, turtles basking in the sun, and dozens of other species from this rich ecosystem.

The Namekagon River Visitor Center is in Trego on US Highway 63, and a short distance up the river (east on Highway 63) on the same side of the river is the trailhead for the Trego Nature Trail, a 2.2-mile round-trip loop trail along the river, popular for hiking and snowshoeing. Just downstream from the center, the river makes a broad southwest-to-northwest bend where a gentle current laps against broad, sandy banks. The site has been host to an ancient Ojibwe village, a resting place for river travelers of old, and today, a popular resort and canoe rental.

Protected Wilderness. The Governor Knowles State Forest is a 55-mile-long riverside zone lying downstream of Danbury, designated as wilderness to provide added protection for the riverway ecosystems. The state forest borders the

2.13 Sandhill cranes take flight in Crex Meadows, a wildlife preserve near the St. Croix River where bird life is abundant. STEVEN ROSSI, ROSSI.PHOTO

sprawling Crex Meadows State Wildlife Area—30,000 acres of restored wetland and brush prairie, accessible via County Road F between Danbury and Grantsburg. Home to an astonishing array of birds, butterflies, reptiles, and amphibians, it is known for its heron rookery and as a haven for migrating waterfowl. The Crex Meadows area lies on the floor of Glacial Lake Grantsburg (Figure 2.13).

Where State Highway 70 crosses the St. Croix River at a Wisconsin Department of Natural Resources (WDNR)-managed campground and landing, there is a trailhead for the Sandrock Cliffs Trail system, named for sandstone exposures along the river and managed by the National Park Service. This system of looping trails provides close views of the river and of ancient riverside cliffs. The cliffs are made of crumbling sandstone deposited 500 million years ago by Cambrian seas and exposed by the same flooding that carved the Dalles of the St. Croix in Interstate Park.

St. Croix Falls and Interstate Park. The St. Croix River National Scenic Riverway Headquarters and Visitor Center is housed in St. Croix Falls. It overlooks the powerful rapids at the head of the Dalles of the St. Croix—a canyon carved from basalt by postglacial floodwaters (Figure 2.10). The canyon exists because mounds of basalt upstream of St. Croix Falls have channeled the flow of every version of the St. Croix River over the centuries, including the catastrophic postglacial floods that carved the gorge. It is accessible via several trails in Interstate State Park, which also take hikers to the park's famous potholes (Figure 2.11) and other important geological features. The trail system and park features are explained well at the park's Ice Age Interpretive Center.[21]

Osceola Bedrock Glades and Cascade Falls. Between Interstate Park and Osceola, County Road S (or River Road) runs roughly parallel to the river. This winding road passes through bedrock glades—rare forest plant communities that have taken hold on thin, rocky soils overlying basalt bedrock. Hikers can explore these glades using the Riverview Trail, which departs from a parking area west of County Road S just north of Osceola. Take care hiking through these ecosystems, for they are rare and fragile, hosting a variety of ferns, cedars, and other hardy species.

Osceola is one of those old river towns that grew up on a bench, or terrace, along the river. Surrounding Osceola is an area of dolomite laid down by invading Ordovician seas over Cambrian sandstone. Having withstood eons of erosion, it is an island of dolomite surrounded by Cambrian bedrock. On the west side of town, Osceola Creek cascades down into a small but impressive canyon notched into the dolomite bedrock. The creek drops over a dolomite ledge and continues to wear away at the softer sandstone beneath it. The 25-foot-high Cascade Falls is a popular hiking destination, accessible via a staircase down the bluff in downtown Osceola.

Hudson and Willow Falls. On the north side of Hudson, County Road A departs to the east, leading to Willow River State Park. Here the Willow River cascades down through a deep dolomite formation and plunges over several wide ledges of more resistant dolomite a few miles upstream of where it empties into the St. Croix (see photo on page viii).[22] The colorfully historic city of Hudson sits on a terrace that was once part of the St. Croix riverbed. Birkmose Park in west-central Hudson provides an elevated view of the St. Croix River and valley. Before the last glacier's meltwaters largely filled the river valley with outwash from the retreating masses of ice, the valley at this point was 210 feet deeper and much wider than it is now, so the scene would have been quite different here thousands of years ago.[23]

Long before the river flowed, the Hudson area was heaved up due to the compression of the crust that created the St. Croix Horst at the end of the

Keweenawan era. Hudson sits on the Hudson-Afton Horst, which itself lies within the St. Croix Horst. This means that the city sits on a three-mile-wide strip of Cambrian sandstone that was raised above the younger Ordovician rock layers on either side of it. The St. Croix River crosses the horst at Hudson and has eroded a channel through it. Exposures of this sandstone on the riverbank are on display in downtown Hudson, where the sandstone layers are nearly 40 feet thick.[24]

The experience of traveling along the river at Hudson is entirely different from that of touring the river's upper reaches. Hudson and a few other smaller towns sit on Lake St. Croix, a natural widening of the river. On summer days, the lake is peppered with every kind of watercraft, and the river towns are all fronted by marinas bristling with the masts of sailboats. Downstream, the river was partially dammed by two masses of sand and other sediments. The first was pushed into the St. Croix over centuries by the Kinnickinnic River, where it meets the St. Croix 12 miles downstream of Hudson. The second is 8 miles farther downstream at Prescott where the St. Croix flows into the Mississippi River. Together, these partial blockages slowed and widened the river.

Prescott and the River's Mouth. The St. Croix River ends its journey flowing placidly into the nation's largest river, the Mississippi. At the confluence, the St. Croix and Mississippi riverbeds are made up of layers of Ordovician dolomite. This rock is overlain by St. Peter sandstone, which makes up the river bluffs and is known for its pure quartz content and striking yellowish-brown outcrops on the bluffs.

The city of Prescott's river walk affords a good view of the confluence and passes through Mercord Mill Park. Above a marina located near the park is a ledge of dolomite, and below that, a beach once existed. It has been underwater since the Mississippi River level was raised here by damming. That beach was long ago the site of a Dakota village. In 1736, a party of Ojibwe warriors paddled down the St. Croix, attacked the village, and defeated its defenders, thus helping to secure their status as the dominant tribe in the area. The Great River Road Visitor and Learning Center, offering a wealth of historical information about the St. Croix and Mississippi Rivers, is perched on a bluff on the south side of Prescott overlooking the confluence of the two rivers.

THE BOIS BRULE RIVER

At the northeast end of Upper St. Croix Lake—the long, narrow body of water that lies next to the town of Solon Springs—is a heavily used boat landing. A few yards from its parking lot, across County Road A, is a trailhead for a trail that runs northeast along a wooded ridge bordering a mile-wide swath of wetland. It is a segment of the increasingly popular North Country National Scenic Trail.

Imagine that we could stand at this parking lot facing northwest—the lake to our left and the trail to our right—on a sunny, cool day in autumn and suddenly turn the clock back three centuries. The lot would vanish and the lake would drain away rapidly to be replaced by a narrow stream flowing south. The large wetland to the north, however, would remain unchanged, and the trail would also remain. It would be narrower and less trampled than it is today, but there it would be, running northeast up the ridge that borders the long wetland.

Now imagine hearing the muffled sounds of voices in the distance to our left and turning to see a small group of men paddling four canoes upstream in our direction. Sunlight flashes off their paddles, and as they come closer, we see that the canoes are made of birch bark and are loaded with bundles wrapped in deer hide and tied with leather straps. The eight men, dressed lightly in animal hides, land their canoes on the sedge-covered low bank of the stream and begin unloading their cargo. They strap the bundles onto their backs, hoist the canoes, and begin packing their loads up the trail to our right.

If we could speak with these men before they depart, we would learn that they are members of a band of Ojibwe transporting goods—kettles, colored cloth, glass beads, axes, and tools, acquired from French traders in exchange for deer, moose, elk, and beaver pelts—from some point on the Mississippi River southwest of us. They are among thousands of Indigenous traders who, every year, ply a vast network of streams and rivers in North America. These men have just made the long passage up the St. Croix River from where it meets the Mississippi and are now on the way to their village located north of us on the shore of Lake Superior. This last leg of their journey begins with the two-mile hike on the portage trail bordering the wetland. At the end of their portage, they will begin perhaps the most challenging part of their entire trip, canoeing their goods down the Bois Brule River (Figure 2.14) to their village on the lakeshore.

On the river, they will first travel a long, winding route in a narrow stream

2.14 The Bois Brule River

and cross a couple of wide stretches of flat water before beginning several miles of a wild ride on one set of frothing rapids after another—dozens in all. The end of their journey will be more restful as the river flattens out, slows, and meanders between high banks of red clay the rest of the way to their destination—the big lake they call *Gichigami*.[25]

The trail referred to here is one of the oldest portage trails in North America—the portage from the headwaters of the St. Croix to the headwaters of the Bois Brule. It lies on a high ridge bordering the wetland where the two rivers rise (Figure 2.15). It traverses a topographic divide separating streams that flow north to Lake Superior from those that flow south and southwest toward the Mississippi River. If we could follow these early Ojibwe traders, we would be taking a trip down one of North America's best-known rivers—cherished over the

2.15 The vast bog where both the Brule River (pictured) and the St. Croix River originate. GAIL MARTINELLI

centuries by residents and visitors who include fly-fishers, canoeists, wealthy landowners, and five US presidents.

The Bois Brule is commonly called "the Brule," not to be confused with the Brule River that forms part of the Michigan–Wisconsin border in far northeastern Wisconsin. The French phrase *Bois Brule* translates to "burning wood." The Ojibwe name for the river is *Wiisaakode-ziibi*, which means "a river through a half-burnt woods."[26] Both names refer to a major fire that burned the valley. Later, British explorers used the name Burntwood River.

The Brule originates in a bog just northeast of Solon Springs and flows northeast for a few miles, then north for the rest of its run until it empties into Lake Superior about 20 miles east of the city of Superior. The river valley is about 30 miles long, but within the valley, the winding river takes a 44-mile course from source to mouth. The valley lies mostly in Keweenawan sandstone, deposited around 1 billion years ago and now exposed at several points along the river banks, especially in the lowermost stretch of the river. However, a five-mile section of the valley crosses a zone of harder, more erosion-resistant Keween-awan basalt, part of a body of older rock mostly buried beneath the sandstone but uplifted about 900 million years ago along the Douglas Fault (see Chapter 1).

In the Brule Valley, the section crossing the basalt has been eroded by glacial ice and flowing water to the point where it is indistinguishable to the untrained eye from the rest of the riverbed lying in sandstone. While the Douglas Fault has created spectacular waterfalls on other rivers, the Brule tumbles over the fault in turbulent rapids where water and ice have been chipping away at the hard basalt bedrock for millions of years. Another name for basalt is trap rock, and this section of the Brule Valley is sometimes called the Trap Range. The lava that formed the basalt rock lifted copper from the depths of the earth's mantle to near its surface. These copper deposits, among the purest in the world, explain why copper mining became a major industry along the Lake Superior shores. The Trap Range in the Brule Valley has also been called the Copper Range.

Upstream and downstream from this range of hard bedrock and rapids, the riverbed lies in sandstone, and the Brule flows more easily with scattered, gentler rapids. In its lowermost segment, the river flows through deep deposits of reddish sand and clay that blanket much of the Lake Superior lowland. On the 18-mile stretch between the Trap Range and the river's mouth, the Brule drops 328 feet, but the last stretch to the sprawling mouth of the river is flat water.

The valley of the Brule was formed, beginning around 2.5 million years ago, by both the infinitesimally slow flow of glacial ice and by the infinitely faster flows of glacial meltwaters. Each advance of the several glaciers scraped away more sandstone and plucked any loosened pieces of the harder basalt and moved them, often several miles, before dropping them wherever the ice halted before melting back. In this way, ice slowly carved a valley that would carry great floods of meltwater from the receding glacier.

As the glacier shrank away to the north, meltwater pooled to form Glacial Lake Duluth, which occupied portions of the Lake Superior basin for centuries, its north and northeast sides lapping against the shrinking wall of ice. That lake's water level got as high as 500 feet above the level of today's Lake Superior. This put its south shore close to a low, rough ridge of basalt raised by the Douglas Fault, which lay parallel to the lakeshore a few miles south of present-day Highway 2. That ridge had a low point where the Brule Valley now lies, and the entire area was covered by hummocky moraine material sheathed in ice that formed a fragile dam holding back Glacial Lake Duluth.

As temperatures warmed and lake levels rose, the dam's days were numbered. At some time between 10,000 and 9,500 years ago, it succumbed to rising lake waters, which then coursed through the breached dam at the low point on the basalt ridge. The resulting massive flood surged south across the headwaters area and into the valley of the ancient St. Croix River. This flooding continued until the level of Glacial Lake Duluth dropped to a certain point, uncovering north-sloping land where the two streams that make up the Brule's headwaters began to flow north again, cutting through thick beds of clay and combining to form today's Brule River.

This sequence of events probably happened several times during the Ice Age. However, the evidence studied by geologists suggests that today's Brule River Valley was formed mostly by the Wisconsin glaciation. For example, along the banks of the Brule closer to its mouth are signs of four different lake levels higher than today's level—old gravelly shorelines and beaches set into the clayey banks. Geologists' careful measurements have revealed that some old shorelines are not level, providing evidence of glacial rebound that has tilted the land slightly—an effect attributable only to the most recent glaciation.[27]

As the postglacial flooding diminished over time, the Brule slowly carved a narrower channel within its broad deposits of sand and gravel, leaving a terrace on one or both sides of the river. This process repeated itself over time, creating a series of terraces, one below the next. The youngest terrace is today's swampy floodplain of the valley, while the older, higher terraces eventually become forested. On the Brule, especially near its headwaters, the next-youngest terrace is about 100 feet above the floodplain, the next about 40 feet higher, and the oldest about 20 feet higher still.[28] They indicate stream levels of the past, which in turn provide another sign of the changing levels of ancestral versions of Lake Superior.

The nature of the Brule's headwaters area has played a key role in making the river attractive for human use and habitation for centuries. The river originates in a remote bog that is 1 mile wide and 10 miles long (Figure 2.15). It lies in a vast sand barrens, 5 to 15 miles wide and more than 100 miles long. Together, the barrens and bog cleanse and store water, providing the area's streams and rivers with a steady supply, even in times of drought. The Brule begins as two small streams fed by this bog that converge to form a six- to eight-foot-wide stream that winds for nine miles through the wetland among alders, cedars, tamarack, and spruce. The surrounding sand barrens collect and filter precipitation, which flows gently downslope into the bog and emerges as a set of prolific springs, described as follows in a 1930 surveyor's report to Congress: "These springs vary in area from one-quarter of an acre to three acres, and are about 5 or 6 feet in depth. In the bottom of most of the . . . springs there may be seen, boiling up through the sand gravel, jets of water varying from one inch to as much as 10 feet in diameter."[29]

This image suggests a steady supply of clear, cold, well-filtered water surging steadily down the river. Such waters, flowing over an alternately sandy and rocky riverbed, provide ideal habitat for trout—and for anglers who consider the Brule to be one of the finest trout streams in the upper Midwest. Brook trout are native to the river, and other species have been introduced by the WDNR. Brown and rainbow trout and Coho and Chinook salmon migrate upstream from Lake Superior during part of the year.

The river's reliable flow made it an ideal route for the canoes of American Indian and French traders of old, who traversed the river in both directions. In recent decades, the river has been a popular route for modern-day recreationists. Its long stretch of almost continuous rapids above and below the Trap Range—especially two sets of basalt ledges (Figure 2.16)—make an exciting and challenging ride for whitewater paddlers. The river valley is also popular for bird-watchers, who have identified more than 200 species there. Rare species include the black-backed woodpecker, white-winged crossbill, merlin, great gray owl, and goshawk.

The story of human habitation of the Brule Valley probably begins with Paleo-tradition hunters of big game who followed the margin of the glacier as it shrank to the north around 12,000 years ago. This is indicated by arrowheads found near Solon Springs. Other archeological evidence suggests that groups of 15 to 60 people moved seasonally from camp to camp, the hunters among them seeking woodland caribou, deer, bear, and elk. The Brule River Valley was fertile ground for hunters, hosting herds of caribou that are estimated to have numbered between 100 and 1,000.[30]

During the Archaic period (3,000 to 8,000 years ago), the Old Copper culture arose. Native people mined shallow deposits of copper and used the mineral to make fish hooks, ax heads, knives, arrowheads, awls, and decorative ornaments. These people were part of a vast trading network that stretched from Lake Superior to the Gulf of Mexico, so it is thought that they could have been the first people to use the Brule–St. Croix Portage Trail, which would have been an important link in this network.[31]

2.16 The Bois Brule River flows over resistant basalt on the May Ledges where many canoes have capsized.

Between 3,000 and 600 years ago, a group of Woodland culture people known as the Laurels occupied much of the Brule Valley. They, too, were copper miners and hunter-gatherers. They built burial mounds in the valleys of the Brule and St. Croix and plied both rivers in dugout canoes. They made syrup and sugar from the sap of maple trees. They were succeeded by more hunter-gatherer tribes, the Potawatomi and Mascoutins, and still later by other tribes—the Dakota, Odawa, Huron, and Ojibwe. Eventually, the Ojibwe became the dominant tribe in the valley. On the upper stretch of the Brule, halfway from its source to Stone's Bridge, where most modern-day paddlers put in, is a clearing referred to in guidebooks as Second Lunch Ground, which is thought to be the site of an Ojibwe hunters' camp.[32]

By historical accounts, the Native American trade network evolved and worked well for thousands of years, until French explorers ushered in the fur trade, which disrupted the Native trading ecosystem and led to its demise. Its effects are thought to have reached the Brule River Valley in the 1650s when French voyageurs began treading the ancient Brule–St. Croix Portage Trail.[33] The Ojibwe were enticed by the French explorers to take part in the burgeoning fur trade. In addition to kettles, metal tools, and other conveniences brought by the newcomers, the Ojibwe also gleaned guns and ammunition, which gave them a decided advantage in their battles with competing nations.[34]

The Ojibwe fought for decades with the Dakota for control of the Brule River Valley. In 1842, a group of Dakota Indians attacked an Ojibwe camp near the mouth of the river and were badly beaten in what became known as the Battle of the Brule—a decisive victory in the Ojibwe quest for control of the region. This was less than a decade before the US government forced the Ojibwe, along with other tribes, onto reservations after the Treaties of La Pointe, negotiated in 1850.[35]

The Brule River sustained many generations of Indigenous people without suffering serious degradation. In the late 1800s and early 1900s, the valley survived short-lived, sporadic copper mining and farming efforts, and many trees were cut down during the lumber boom of that same period. But by the 1930s, these enterprises had been abandoned and the valley's land had been left to heal. In 1933, the Civilian Conservation Corps built a camp near the town of Brule and worked on fire control and reforestation over the following eight years. Today, the ecosystem is somewhat challenged by the annual influx of anglers and recreationists, but a deep forest once again shelters the riverbanks, and the water still flows cold and clear for the length of the river.

The Brule is highly revered by landowners and by most visitors. Many who own land along the river are vocal about protecting it and keeping it pristine and safe from overuse by visitors. Some have worked with conservationists to protect the river's fish populations from pollution and overfishing. Among the many stately lodges on the river, the most famous property is the Cedar Island Estate. Its sprawling lodge, outbuildings, and fishponds, not accessible to the public, lie along a quiet stretch of the upper Brule where several famous Americans, including generals and industrialists, have fished. Over the years, the lodge has hosted five prominent fishermen who were or would later become US presidents: Ulysses S. Grant, Grover Cleveland, Calvin Coolidge, Herbert Hoover, and Dwight D. Eisenhower.[36]

The Brule River State Forest was established in 1907 on 3,000 acres of land donated to the state. As of 2008, it had expanded to 41,000 acres, encompassing the entire length of the river (although 12,000 acres within the valley remain privately owned).[37] With rich, relatively undisturbed soils in some areas, the forest hosts rare old stands of sugar maple and hemlock. The state forest contains a 16-mile stretch of the North Country National Scenic Trail and eight miles of Lake Superior shoreline.

> ## TRAVEL GUIDE
> ### *The Bois Brule River*
>
> Paddling the Brule is by far the best way to experience it, although its famous rapids make this a challenging way to go. You can see many of the Brule's extraordinary features by driving through the valley and hiking some trails along its banks. A good way to begin is to hike the ancient Brule–St. Croix Portage Trail. The heavily wooded trail is mostly flat, except for a few steep sections. It lies on a high ridge overlooking the wetland of the Brule and St. Croix Rivers' headwaters, although views of this valley are limited because of the density of the forest. On the trail, hikers follow a path trod by many generations of Indigenous people as well as hundreds of early European explorers and fur traders. Hikers can take three spur trails to get views of the St. Croix River headwaters, the headwaters of the Brule, and the point where travelers entered or exited the Brule River for centuries, finishing or beginning their portage between watersheds (Figure 2.15).
>
> To continue the drive up through the valley, starting from the County Road A boat landing, go south a short distance on A and turn left onto Jerseth Road, which leads to County Road S. Take a left on S and follow it less than two miles to Stone's Bridge Landing, a popular put-in site for Brule River paddlers. Paddling, or just looking upstream from the bridge, you can get a good feel for the tightly winding nature of the Upper Brule.
>
> Follow Highway S north about five miles to the community of Lake Nebagamon and the junction with County Road B. Take B east for four miles to where it crosses the Brule and meets State Highway 27 at Winneboujou, another

access point for canoeists. Just downstream of this bridge begins one of the long stretches of wild rapids that make the river famous. From Winneboujou, Highway 27 skirts the river valley on its east side, running north to the town of Brule, where it ends at US Highway 2.

For an alternate route from Winneboujou on the west side of the river, take Congdon Road north for two miles to where it ends at Ranger Road, which continues north to Brule. Ranger Road leads to the Bois Brule Campground and Brule River State Forest headquarters where you can find a wealth of information about the history of the Brule and about how to use the river sustainably. The Stoney Hill Nature Trail out of the Brule River Campground is a 1.7-mile loop that takes hikers to views of the entire river valley, which was once filled to the brim with the raging waters of the postglacial flood.

From Brule, take County Road H north for a little more than five miles to its junction with County Road FF, which runs west to the river. Angler parking areas are located in several spots along Highway H. These access points allow closer looks at the river, which is docile and winding for much of this stretch. County Road FF crosses the river on the Johnson Bridge, and just upstream of that point, the river becomes challenging for boaters and even hazardous in high water. Early on this stretch are the Lenroot Ledges—four sharp drops in a set of ferocious rapids that have challenged many a canoeist (including, several times, yours truly). They lie in the Copper Range, the stretch where the river crosses basalt bedrock and is anything but docile.

Below the Johnson Bridge, a fast half mile takes canoeists and kayakers to another challenging set of sharp drops called the May Ledges (Figure 2.16). They can be viewed from the east side of the river, accessible via Koski Road, which leaves Highway H less than a mile north of its junction with FF. Here the river has crossed the Copper Range and, after another short stretch of medium-hazard rapids, settles into an easier flow as it works its way through deep red clay and sand laid down upon ancient sandstone by Glacial Lake Duluth. For canoeists, the rest of the ride is easier, with several medium- and low-hazard rapids. For drivers, the next access point is north about three miles on Highway H, then west on State Highway 13 to a wayside on the river.

From its junction with H, Highway 13 runs north two miles to Hoefling Drive, which goes to McNeil Landing on the river (for paddlers, the landing is three river miles downstream of Highway 13). North of here, much of the valley is encompassed by the Brule River Boreal Forest State Natural Area. On the terraces

2.17 The mouth of the Bois Brule River on Lake Superior's south shore

and steep slopes bordering the river grows a northern forest—featuring large white pine, white spruce, balsam fir, balsam poplar, and white cedar—which has been recovering since loggers leveled it in the early 1900s. The wet terraces support swamp hardwood stands—black ash, red maple, alder thickets, and marsh areas. Bird-watchers look for warblers (black-throated green, pine, and Nashville), ovenbirds, and hermit thrush. This state natural area, accessible via fishing access points, has preserved a new generation of the ancient forest that stood for so long—an environment in which you can pause to appreciate the long-evolved natural beauty of the Brule River Valley.

For the last leg of the drive to the river's mouth, head north another three miles on Highway 13 to Brule River Road and turn left to follow that road along the river to the boat landing on the Lake Superior shore. At the end of its journey, the river is sluggish, taking a lazy turn to the west to empty its tannin-stained water into the big lake (Figure 2.17). Here you can hunt for agates, listen to the gurgle of river water flowing across the sprawling sandy delta, and contemplate the breathtaking view of the vast inland sea to the north.

THE CHIPPEWA RIVER AND ITS TRIBUTARIES

All of northern Wisconsin was shaped by a crushing field of ice that spread over the land for millennia, flattening peaks and ridges and filling hollows and valleys with rock and soil. As it melted away 10,000 years ago, it draped the land with thick layers of silt, sand, and gravel that were eventually shaped into ridges, peaks, hollows, and valleys by voluminous meltwaters. Thick vegetation evolved on this new landscape as countless rivulets trickled from boggy headwaters nestled in deep woods, winding among greening hummocks and through lush valleys and hollows. These little streams merged to form bigger streams that in turn merged to form rivers in a complex network that now drains a land loaded with water by the glacier—a key part of Earth's ancient and ongoing water cycle. The Chippewa River (Figure 2.18) and its tributaries form a classic example of such a watery network. They drain nearly one-fifth of Wisconsin's land area, ultimately emptying all that water into the Mississippi River.

The watershed drained by the Chippewa and its tributaries is splayed across northwestern Wisconsin, shaped like an evergreen tree with its trunk stuck

2.18 The Chippewa River near Jim Falls

into the Mississippi Valley, its axis pointing to the northeast, and its topmost branches brushing the border between Wisconsin and Upper Michigan (Figure 2.1). The northeast half of the basin, referred to as the Upper Chippewa River Valley, lies on Precambrian igneous and metamorphic bedrock. It contains some of the oldest rock outcroppings in the state, including 1.6-billion-year-old quartzite exposures in the Blue Hills of Barron, Rusk, and Sawyer Counties, and other outcroppings of volcanic and metamorphic rock around 2 billion years old.[38] The Upper Chippewa Valley also holds remnant features formed by glaciers that preceded the most recent one, at least 790,000 years ago.[39]

In the southwest half of the basin—the Lower Chippewa Valley lying between Chippewa County and the Mississippi—that Precambrian bedrock slopes to the southwest at about 15 vertical feet per mile. Although it is overlain by Cambrian sandstone, it outcrops at several points in the river valley. At the mouth of the Chippewa, however, because of the slope, the Precambrian rock is buried 500 feet below the land surface, under layers of sandstone and spotty deposits of Ordovician dolomite. The dolomite deposits now cap many of the high ridges of the Driftless Area through which the Lower Chippewa flows.[40]

As the ice of the Wisconsin glaciation inched across the state, the Chippewa Lobe reached its farthest extent about 18,000 years ago—a margin now defined by a line running southwest across Lincoln and Taylor Counties and curving east and then northwest across Chippewa County. The thickest, central part of the lobe was channeled south by the ancient Chippewa River Valley. As it melted away, the Chippewa Lobe left a vast, deep field of debris that formed hummocky terrain lying north of the broad Chippewa Moraine. The mass of sand, gravel, and boulders that the ice had scraped together over many centuries was dropped on the land to form a random arrangement of hills (hummocks), ridges, valleys, and hollows. Stranded massive chunks of ice that had calved from the glacier slowly melted to form kettle lakes.

At the southwest end of the Chippewa Lobe, the Blue Hills, made of ancient quartzite, had resisted the invading ice and were not nearly as eroded as the surrounding land. However, because the ridges and valleys of the Blue Hills trend northeast to southwest—the direction of glacial movement in the valley—geologists can surmise that the glacier did carve them to some extent. The southwest end of this 20-mile range of hills is steeper than the northeast end, indicating that the glacier slowed and thinned when it reached that area. Southeast of these hills is a high ridge of quartzite—the Flambeau Ridge—formed by the same process that made the Blue Hills. This ridge would play a major role in the formation of the Chippewa Valley in postglacial centuries.

The headwaters of the Chippewa originated as subglacial streams gushing from beneath the glacier as its mass of ice melted. Flowing away from the ice, these waters formed broad, shallow, braided streams flowing in multiple criss-crossing channels that carried sand and gravel from the retreating glacier and spread it out in coarse, thick layers, called outwash, across the river valley floor. The meltwaters of the Chippewa Lobe carried one of the biggest flows of out-wash in the state.[41] Because the waters flowed so hard and fast, broad, sandy, gravelly plains were formed along the river from the Chippewa Lobe's terminal moraine in Chippewa County all the way to the mouth of the river.

The ice eventually dwindled to the north and the broad meltwater flow nar-rowed, incising a smaller channel into the wide riverbed, leaving broad terraces lying alternately on one or both sides of the new channel in the Lower Chippewa Valley. The largest of these is the Wissota Terrace, named for Lake Wissota (site of Lake Wissota State Park) created by the dam north of the city of Chippewa Falls.[42] That terrace was created about 10,000 years ago and was broadest at Lake Wissota, where it now lies underwater. The river's present-day channel was created in stages. Geologists have found a descending series of seven terraces in the valley, created between 10,000 and 2,000 years ago (Figure 2.19).[43]

As a result of this geologic history, the Chippewa varies greatly between its upper and lower sections. Flowing through the hummocky terrain north of the Chippewa Moraine, the river meanders among the hills. As the postglacial river carved its winding valley south across what is now Rusk County, the high ridge of quartzite called the Flambeau Ridge stood in its way (Figure 2.1). To get around this bulwark of hard rock that had withstood the force of the glacier, the river had to veer east at what is now called Big Bend, to where it could again flow south. Beyond that point, the Upper Chippewa keeps winding among glaciated terrain until it reaches the cities of Chippewa Falls and Eau Claire.

The Lower Chippewa—that segment between Eau Claire and the river's mouth—is generally wider than the Upper Chippewa (above Eau Claire) and flows more directly to the river's confluence with the Mississippi River. On the way, it passes first through an area that was glaciated, but not by the most recent glacier, and then through Wisconsin's Driftless Area, which was never glaciated.

Around 10,000 years ago, the ancestral Chippewa River was carrying gla-cial meltwater to the Mississippi, moving a huge load of sand and gravel. At its

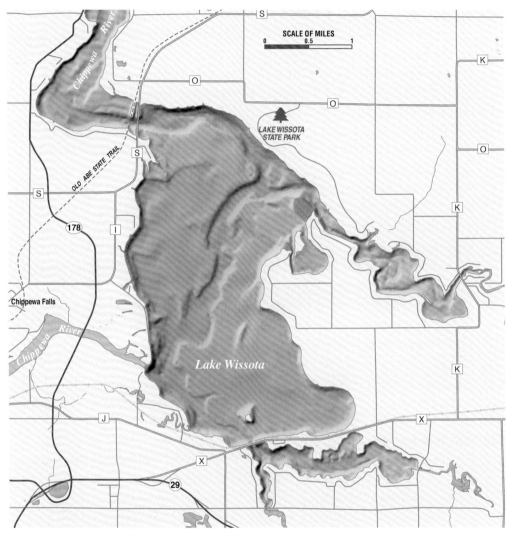

2.19 This map of Lake Wissota shows the deep channel of the Chippewa River weaving through the broad, flat terraces created by the river long before the area was flooded by a dam.
MAPPING SPECIALISTS, LTD., FITCHBURG, WI

mouth, the Chippewa's gradient is slightly steeper than that of the Mississippi. To put it more simply, the river near its mouth flows down a gentle slope and splashes onto the relatively flat Mississippi riverbed, where it slows. In post-glacial days, this caused the ancient Chippewa to drop its load of sand and gravel at the confluence. The resulting mass of glacial debris formed a delta, which partially dammed the Mississippi, enough to create Lake Pepin—the long, wide segment of the Mississippi lying upstream from the mouth of the Chippewa. Even today, the river continues to drop sediments on this delta.

Several centuries after the glacial ice retreated far to the north, the basin of the Chippewa with its intricate network of streams was established. At the head-waters, two prominent streams joined to form the main river. The East Fork was formed from several capillary creeks that still today drain the bogs of southeast Iron County. The West Fork flowed from a small kettle lake in southeast Bayfield County called Chippewa Lake—not to be confused with the Chippewa Flowage, also called Lake Chippewa. The latter is the sprawling flowage created by the damming of the Chippewa River where the two forks meet, just northwest of the town of Winter (see "Lake Country" in this chapter). Beyond the dam, the river flows 185 miles within a wide, deep trench, which was carved by meltwaters over a period of about 500 years.

The Chippewa River Valley has been described as two distinct ecosystems. The change in vegetation types between the upper and lower valleys reflects the different types of soil and bedrock underlying them. The bedrock in the upper valley comprises older varieties of granite, basalt, and other volcanic and metamorphic rock, while the Lower Chippewa flows on Cambrian sandstone. Soils are derived largely from bedrock, but the glaciers also influenced soil for-mation by covering bedrock and older soils with thick layers of sand and silt. The Chippewa Lobe of the most recent glacier stopped in present-day Chippewa County, north of Chippewa Falls. Hence, the soils north of the city tend to be loamier and siltier than those to the south—a result of bedrock differences and the influence of glaciers.

The vegetation that developed in postglacial centuries reflected these differ-ences in bedrock and soils between the upper and lower valleys. Forests of maple, hemlock, and yellow birch (called northern mesic forests) eventually covered the upper reaches of the Chippewa Valley. Around the future site of Chippewa Falls, a pine barren hosting jack pine and prairie grasses formed a transition between the northern mesic forest of the Upper Valley and the prairie and oak savanna of the Lower Valley. On its lowermost 15 miles, the river flows through a lowland, or bottomland, hardwood forest of willow, maple, and ash.

The first people arrived in the Chippewa Valley around 11,000 years ago, and by the time Europeans began exploring the valley in the 1600s, the domi-nant Native tribe there was the Ojibwe. The first permanent Ojibwe settlement was on Lac Court Orielles, a few miles west of the Chippewa, and from there, they expanded south.[44] The Chippewa Valley proved to be a fertile ground for hunting and trapping, which sustained the Ojibwe people during fall, winter, and early spring, and they also fished the area's streams and lakes year-round. Spring brought sap to the maples for making syrup and sugar. The forest provided ample berries and nuts during summer and fall, and wild rice in some of the lakes and wetlands was a staple.

The Chippewa River became a vitally important waterway for the Ojibwe, who established several villages along the river. While moving into the area, they had competed and fought with the Dakota. Between these two tribes, several large villages were established by 1000 CE.

The Ojibwe took an active role in the fur trade, guiding and supplying European traders. Beginning in the mid-1600s, the Chippewa River served as an important link in the network of streams and rivers connecting Lake Superior to the Mississippi River. By the late 1700s, the Ojibwe controlled the Upper Chippewa Valley as far south as today's Chippewa Falls, where they had a number of villages, and the Dakota controlled the Lower Valley. The river is named for the Ojibwe. The word *Chippewa* is thought to have been coined by self-taught American mapmaker Jonathan Carver who explored the river in 1767 and apparently contorted the word *Ojibwe* to create the new name. Until that time, the river had been known as the Buffalo River, so named for the large herds of bison that lived along the lower stretches of the river in those days.[45]

In 1837, the US government coerced the Ojibwe groups living in the area to cede the entire Chippewa Valley and then some by signing the Treaty of St. Peters. The US government, motivated by the desire to control vast tracts of timber, gained control of all the land from the Mississippi to the Wisconsin River and from Lake Superior to Prairie du Chien. According to the treaty, the United States was to make payments to the signatory bands for 20 years, totaling $810,000 "in goods and money," while the Ojibwe retained the right to hunt, fish, and gather on the ceded lands and waters in perpetuity.[46]

Thus began a new era for the Chippewa Valley, which was eventually populated by European settlers. The river became a major conduit for timber taken from the vast valley during the logging era. Between the 1840s and 1970s, for purposes of logging, steamboat travel, flood management, and generation of electricity, an estimated 148 dams were built on the Chippewa.[47] Today, fewer than a dozen are still in operation, impounding water in flowages of varying

2.20 This view of the Chippewa River, accessible from the Chippewa River State Trail, shows the river in its natural state.

sizes, including the Chippewa, Radisson, and Holcomb Flowages; Old Abe Lake; Lake Wissota; and Dell's Pond.

The Lower Chippewa River, downstream of Eau Claire, remains free of dams, and much of it resembles the valley before Europeans arrived (Figure 2.20). According to the WDNR, the Lower Chippewa Valley features "remnant prairie, oak savanna, woodlands, wetlands, and . . . the greatest concentration of rare species of any similar sized area in Wisconsin," including 50 percent of all the state's native plant species. Before European immigration, native prairie covered 7.7 million acres in the state; today it occupies about 8,000 acres, of which 2,000 acres lie in the Lower Chippewa Valley.[48]

The lowermost valley, between Durand and the mouth of the river, also contains the largest intact floodplain, or bottomland, hardwood forest in the upper

Midwest. Lying mostly within the Tiffany Wildlife Area, it has been designated a National Natural Landmark. It hosts a large heron rookery and populations of 70 percent of the state's fish species, with 18 rare species, including the paddlefish, crystal darter, and blue sucker. The area provides habitat for dozens of bird species, including six threatened species: Acadian flycatchers, hooded warblers, Kentucky warblers, red-shouldered hawks, yellow-crowned night herons, and cerulean warblers.[49]

TRAVEL GUIDE
The Chippewa River

Unlike other rivers in the state, the Chippewa is not easily followed by highway. Rather than describe a car tour of the valley, I have highlighted six shorter trips to sites and segments of the river.

Winter Dam to Confluence with Flambeau. Author Richard D. Cornell, who spent years canoeing the full length of the Chippewa, found the stretch of the river between the Winter Dam and the confluence of the Flambeau and Chippewa Rivers to be the best place from which to see "the true untamed river."[50] If canoeing is not an option, several roads can take you close to the river and to several boat landings where you can get the true flavor of the Upper Chippewa. From the dam northwest of Winter heading south, those roads are Dam Road, County Road G to Ojibwa, State Highway 27 to Radisson, State Highway 40 to Bruce, and one mile on US Highway 8 in Bruce to County Road E, which runs south to the confluence.

Flambeau Ridge. Just south of the confluence of the Flambeau and Chippewa Rivers is a two-mile long, 300-foot high outcropping of quartzite called the Flambeau Ridge, a southeastern outlier of the Blue Hills. The ridge and hills were formed around the same time, and they are made of much the same material as Rib Mountain in Marathon County and the Baraboo Hills of south-central Wisconsin. Precambrian seas deposited great amounts of sand that were buried deeply and, around 1.6 billion years ago, metamorphosed into the hard rock quartzite.[51]

These quartzite hills and ridges withstood hundreds of millions of years of erosion by rain, wind, frost action, and glacial ice, while land around them was steadily eroded. We know that glaciers covered the Flambeau Ridge because glacial erratics, boulders carried by the glacier, now sit on its summit. The Chippewa River, even with the added flow of the Flambeau River, its largest tributary, was no match for the Flambeau Ridge and had to veer east to get around it. This segment of the river is called Big Bend. You can view Flambeau Ridge from a boat ramp on County Road D, which runs east-west along the river at Big Bend.

Old Abe Scenic Trail. This trail is an old railroad corridor that follows the Chippewa River between Brunet Island State Park in Cornell and Phoenix Park in downtown Eau Claire. An eight-mile segment of the trail runs along an undeveloped bank of the river and through the town of Jim Falls. Just north of town, the old Cobban Bridge, now closed to all but foot traffic, sits over rapids where some of the oldest rock in the state is exposed (Figure 2.18). Called banded amphibolite gneiss, a metamorphosed form of granite, this rock is thought to have been formed as a result of three different episodes involving continental collisions, between 2.5 billion and 1.7 billion years ago. Thus, the rocks you see in the rapids under this bridge could be up to 2.5 billion years old.[52]

Chippewa Falls and Eau Claire. The river flows through Chippewa Falls and winds its way a few miles southwest to Eau Claire. Both cities host interesting historical sites. In Chippewa Falls, the hydroelectric dam there has put the falls for which the city is named underwater. Just downstream of the dam on a broad terrace on the north side of the river is Chippewa Riverside Park. Northeast of this park is a high sandstone bluff carved by postglacial floods. The terrace and bluff were once occupied by Ojibwe villages. On top of the bluff, the Chippewa County Museum is located in one of the old buildings that now command views of the river valley.

In Eau Claire, the river flows into Dells Pond, created for storing logs in the late 1800s behind the lowermost dam on the river. The Dells Dam was built in 1880 at a point where the river narrows between two high sandstone walls, also called dells, after which the dam and the pond are named. Just upstream of the dam, the river curves from south to north and then south again. The second of these hairpin turns nearly encircles a sandstone mass that rises 80 feet from the normal water level—a mound called Mount Simon, site of a city park. The sandstone here was deposited by an ancient sea upon Precambrian sandstone more than 500 million years ago, making it the oldest type of Cambrian sandstone. Geologists chose this as the reference location for this type of stone, now called Mount Simon sandstone. It outcrops on the north end of this bluff, although most of it was submerged by the damming of the river. The mound provides views of Dells Pond and the surrounding area.

Another striking feature created by the Chippewa River is Halfmoon Lake, downstream of Dells Pond. Here the ancient river meandered sharply north and south, creating another hairpin turn, but many centuries ago, the river eroded a shortcut, abandoning the meander and creating an island. This is the site of popular Carson Park, which hosts a first-rate baseball field (where the great Hank Aaron began his professional career) and quiet wooded areas. It is a fitting place for the Chippewa Valley Museum, an excellent repository for historical documents and other materials that tell the story of the valley.

A few blocks east of Carson Park is Phoenix Park, the pride of downtown Eau Claire—a riverside green space hosting walking trails, a community gathering

2.21 The University of Wisconsin–Eau Claire (the author's alma mater) straddles a terrace on the Chippewa River. UWEC PHOTOGRAPHY

place, and a performing arts facility. This park preserves the confluence of the Eau Claire and Chippewa Rivers, the former being an ecologically and historically important tributary. Shortly downstream, the river bends from south to west where the University of Wisconsin–Eau Claire campus is located (Figure 2.21).

Some authors argue that Eau Claire is really the place where the Chippewa crosses the line between the Upper and Lower Valley designations, changing from one type of river to another. For a much more complete exploration of the entire river, I recommend *The Chippewa: Biography of a Wisconsin Waterway* by Richard D. Cornell. About the change in the nature of the river, he wrote:

> Below Eau Claire, the river's last sixty miles have been set free. It has flowed past eight towns, spun turbines in eight power plants, embraced many tributaries, and now on its way to the Mississippi, it runs unimpeded, sliding past many islands and carrying tons of sand to the big river.[53]

Chippewa River State Trail. One way to see the Lower Chippewa Valley is to hike or bike the Chippewa River State Trail, which starts at Phoenix Park in Eau Claire and follows the river all the way to Durand, about 15 miles upstream of the river's mouth. The trail is a former railroad corridor that traverses diverse landscapes, including river bluffs, prairies, wetlands, and river bottomland. It also has several access points for those wishing to canoe or kayak the river (Figure 2.20).

Tiffany Wildlife Area and Nelson-Trevino Bottoms State Natural Area. The largely undisturbed tract of land in the lower-most Chippewa Valley is occupied by the Tiffany Wildlife Area, a 13,000-acre preserve lying in Buffalo County, east of the Chippewa, and in Pepin County to the west. This is the largest bottomland hardwood forest in the upper Midwest. It is open for hiking and cross-country skiing, as well as for camping (with a permit required), hunting, trapping, and the gathering of wild edibles. There are several access points with parking areas, including the junction of Thibodeau Road and State Highway 25 a few miles south of Durand where an unnamed trail departs to run south along the river to a wayside on State Highway 35. On the west side of the river, parking areas can be found at the ends of several roads that run toward the river from County Highway N, including, from south to north: 16th Creek Road, Tulip Lane, and Swede Ramble.

The Tiffany Wildlife Area is adjacent to the Nelson-Trevino Bottoms State Natural Area, a segment of the Upper Mississippi River Wildlife and Fish Refuge, owned by the US Fish and Wildlife Service. It is open for paddling and hiking via parking areas along Highway 25 south of Nelson. This is a vast bottomland of the Mississippi, created by the natural damming of the river that formed Lake Pepin as the Chippewa deposited its massive load of sand onto the bed of the Mississippi River.

The Upper Wisconsin River

A geographer's description of the Wisconsin River might make it seem like a docile giant: a 430-mile stream that rises in Wisconsin's deep north woods, meanders southward across most of the state's length, then swings westward to flow across the ancient Driftless Area, emptying quietly into the Mississippi River. It drops just 1,067 feet, or two and a half feet per mile on average, between its source and its mouth.[54] The river may, indeed, be docile, but the long, wide Wisconsin River Valley belies a history of upheaval, including continental collisions, a great rift in the planet's crust, crushing glaciers, and catastrophic floods.

These events and processes all played roles in shaping the vast Wisconsin River watershed that now collects water on a fifth of the state's land and channels it into the old river flowing from one state border to another. Because of major differences in the geography and geology of the river's two major stretches, the Wisconsin is really two rivers—the Upper Wisconsin, lying between the river's source on Wisconsin's northern border and the city of Portage, and the Lower Wisconsin (covered in Chapter 5), running between Portage and the Mississippi River.

The watershed that feeds the Upper Wisconsin is spread among 20 counties and includes six major tributaries—the Tomahawk, Rib, Eau Claire, Big Eau Pleine, Yellow, and Lemonweir Rivers—and hundreds of smaller streams. Geologists have found evidence of major geologic events and processes that shaped the Upper Wisconsin River Valley. For example, rock exposures in some of the river's rapids reveal the base remnants of a major mountain range—the Penokean Mountains—uplifted around 1.8 billion years ago in the collision of two primitive continents. And the overall slope of the land that sends the river southward is due to the creation of the Wisconsin Dome, itself a result of the Midcontinent Rift around 1.1 billion years ago.

2.22 The Upper Wisconsin River

Another major player in the shaping of the Wisconsin River Valley was millions of years of erosion by flowing water, wind, and frost action after the last of several ancient inland seas departed the region. During Cambrian and later periods, a series of such oceans covered the region on and off over a period of more than 200 million years. On top of Precambrian basement rock, made up of granites and other crystalline rock types, these seas and the rivers flowing to them deposited thick layers of sedimentary rock—sandstone during Cambrian time and dolomite during succeeding geologic periods. After the last of these seas had withdrawn more than 300 million years ago, erosion carried away most of the upper layers on the Wisconsin Dome, exposing Cambrian sandstone into which certain rivers carved their valleys. The Upper Wisconsin River etched its valley deeply enough along some stretches to expose the older Precambrian rock layers.

The biggest players in the shaping of the Wisconsin River Valley were the glaciers. Soils in the valley today lie over deep layers of silt, sand, gravel, and boulders carried by the glaciers and dropped as they melted away. The average depth of such glacial debris in the valley is 100 feet, ranging from zero to 240 feet, so the valley of the Upper Wisconsin River is set into a glacially formed landscape.

In the most recent glaciation, the Wisconsin Valley Lobe advanced from the Lake Superior basin into the northernmost part of the Wisconsin River Valley, carving it wider and deeper and building a terminal moraine called the Wood Lake Moraine, which arcs across southern Lincoln County just north of Merrill. As the glacier retreated, beginning about 18,000 years ago, its massive flow of meltwater put the finishing touches on the postglacial northern part of the river valley. Glaciologists have described the ancient Upper Wisconsin as a broad stream of meltwater flowing between walls of ice for centuries before it all thawed.[55]

The fact that the river valley channeled the ice of the Wisconsin Valley Lobe means the valley existed before the most recent glacier, probably for several hundred thousand years. An area south of the Wood Lake Moraine straddling much of Clarke and Marathon Counties is layered with till from one or more earlier glaciers. An earlier version of the Wisconsin Valley Lobe advanced to just north of Wausau. That lobe and its meltwaters created the channel for the river between Merrill and Stevens Point, so in that stretch of the valley, the river now flows just as it has for hundreds of millennia.

South of Stevens Point, a long segment of the river valley lies on unglaciated land but was formed indirectly by the glacier. About 19,000 years ago, some 60 miles south of Stevens Point, the Green Bay Lobe of the glacier crept into the Baraboo Hills from the east, blocking the flow of the ancient Wisconsin River in the vicinity of present-day Wisconsin Dells. The river waters backed up and flooded an area the size of the Great Salt Lake in Utah. At the time, the region probably resembled today's Driftless Area of southwestern Wisconsin, so the icy waters filled deep valleys and rose among sandstone ridges and peaks, the highest of which became islands in the newly formed Glacial Lake Wisconsin.

The lake was roughly heart-shaped, bound by the Driftless Area highland on its southwest side, the ice wall of the Green Bay Lobe on its east side, and the moraines of earlier glaciers on its north side. If the lake existed today, the cities of Black River Falls, Neillsville, and Stevens Point would sit on or near its shores. Wisconsin Rapids, Friendship, and Mauston would all be underwater.

Glacial Lake Wisconsin sat in the glacial deep freeze for 3,000 or more years while its waves and currents, driven by the high winds of glacial times, wore away at the sandstone ridges, buttes, and mounds in the lake. Many of the landforms that stood above the water as islands remained in place where they are today, including Mill Bluff, Long Bluff, Friendship Mound, and Roche-a-Cri.

Underwater structures were whittled and dismantled by the waves and currents, their sand grains spread evenly on the ancient lake bottom to form what today is called the Central Sand Plain. Beneath those deep layers of sand, the ancient Wisconsin River bed is buried. It flowed more directly south than today's channel, which takes the shape of a bow arcing west from Stevens Point, bending south through the heart of the Central Sand Plain, and curving back to the southeast as it approaches Wisconsin Dells.[56]

As the climate began to warm around 14,000 years ago, it was only a matter of time until the ice dams that held Glacial Lake Wisconsin at its south end would give way. When they did, the result was a catastrophic flood that would change the landscape markedly and have an equally dramatic effect on Wisconsin's history right up to the present day.

The flood began when the fragile agglomeration of ice and morainal material sitting between the withering Green Bay Lobe and the quartzite mass of the Baraboo Hills disintegrated to the point where it could no longer restrain Glacial Lake Wisconsin. The tremendous volume of water broke through the dam and must have flowed with incredible fury as it quickly stripped away lake bottom and glacial sediments and began carving up the underlying sandstone. Geologists estimate that within a few days or weeks, the entire great glacial lake had drained away. In the process, it sculpted the famous gorge of the Wisconsin Dells (Figure 2.23), along with similar dramatic features such as the glen at Mirror Lake State Park.

The meltwaters of the glacier flowed for many centuries, and together with the drainage of Glacial Lake Wisconsin, they spread untold volumes of clay, silt, sand, and gravel across the Wisconsin River Valley, everywhere downstream of the Wood Lake Moraine. Early in the process, the postglacial meltwater river was many times wider than it is today, sometimes taking the form of a massive single flow, but more often as a broad braided stream.[57] The wider river first troweled outwash across the valley, and later when the meltwater flow slowed, the river narrowed and bored a newer, deeper channel into the outwash, leaving broad terraces on either side of the river. In some areas now, old terraces sit higher than younger ones closer to the river, because this downcutting, terrace-making process occurred more than once. Some of these terraces now lie underwater behind dams at locations such as Biron, Rothschild, and Mosinee. Others are occupied by towns, railroads, and highways.

Outwash was a major shaper of landscapes, waterways, and natural

2.23 Dells of the Wisconsin River, a canyon in sandstone

communities. One interesting example is the confluence of the Rib and Wisconsin Rivers near Wausau. The Rib River was formed by meltwater draining the glacier once it had advanced to its maximum extent northwest of Wausau. As the glacier shrank away, its flow lessened fairly quickly. However, the Wisconsin River carried far more outwash than did the Rib because it drained a much larger area for a longer time. Thus the Wisconsin's floodplain at the confluence was built higher than that of the Rib and served as a dam. The result is a naturally created lake and broad wetland on the lower Rib River at its confluence with the Wisconsin, which has served as a rich resource for aquatic communities as well as human communities throughout the state's history.

The Wisconsin River has never stopped shaping and reshaping its valley.

The process goes on day and night, season after season, due to the complex machinations of river hydraulics. University of Wisconsin–Madison professor Richard D. Durbin has studied the river as extensively as any scientist or historian and has published his findings in *The Wisconsin River: An Odyssey Through Time and Space*. He summarized the ongoing makeover of the river valley as follows: "The channel shifts position, becomes separated into several courses or reunites; bottom features move and formations like sand bars and willow islands are constantly being created, reshaped, and destroyed. . . . One must remember that the river is in reality a finely tuned, interdependent system in which any change, no matter how trivial, will have repercussions all along its course."[58]

Despite its ever-changing nature, a good description of the Upper Wisconsin can capture its mystique. Perhaps no writer has done a better job of that than the renowned geographer Lawrence Martin in his seminal book *The Physical Geography of Wisconsin*. Martin describes three distinctive stretches of the Upper Wisconsin, each of which, he claims, tells a different story of geology and natural history.

The uppermost stretch of the river begins in Lac Vieux Desert, which straddles the Michigan–Wisconsin border in Vilas County. The lake collects the waters of several small creeks that arise in bogs typical of this region of the state. Between the lake and Merrill, the river ambles among an array of kettle lakes.[59] The landscape is hummocky, and within it lies the steepest segment of the river, a rapids called Grandfather Falls, where the river drops 90 feet in 1.5 miles.[60] With huge, polished boulders and exposures of Precambrian rock (Figure 2.24, left), Grandfather Falls is a striking cascade of rapids, especially in high water (Figure 2.24, right), even though a large volume of the river's flow has been diverted around the falls to power a hydroelectric dam.

The next stretch of the valley, between Merrill and Stevens Point, lies in a region with no lakes as it was glaciated hundreds of thousands of years ago, but not by the most recent glacier. Erosion has eliminated many glacial features, smoothing the landscape. Branching systems of streams have developed and most natural lakes have been drained by these systems or have been converted to wetlands. Both of these northern stretches of the Wisconsin River flow within what is called the Northern Highlands of Wisconsin.

The third segment of the Upper Wisconsin is the 181-mile stretch of river between Stevens Point and Portage—the Central Sand Plain. The river begins its journey across the plain in a short valley in Cambrian sandstone, flowing west

2.24 Grandfather Falls in winter, with Precambrian boulders as big as 6 feet in diameter, and in summer during high water

from Stevens Point on a course flanked by old glacial moraine to the north and the vast Glacial Lake Wisconsin lake bed sprawling to the south. It then veers southwest to amble across the sand plain. Martin describes the long segment of the river between Wisconsin Rapids and Wisconsin Dells as flowing in an "aimless pattern" through a shallow valley with scattered sandstone outcroppings.[61]

The evolution of plant and animal communities in the valley of the Upper Wisconsin reflects the geologic differences among the three segments of the valley. As the glacier retreated, the valley hosted tundra, which slowly gave way to scattered spruce, fir, and jack pine. Soils developed and after thousands of years came tree species that depended on richer soils, including white pine, maple, and eastern hemlock.

By the 1700s, the uppermost stretch of the valley was open and boggy, and the river was a shallow, slow, meandering stream near its source. Throughout the rest of the Northern Highlands segment of the valley, the watershed was dominated by hemlock, sugar maple, yellow birch, and white and red pine forests, with smaller stands of jack pine barrens and oak forest. In the part of the region that had been glaciated (but not by the most recent glacier), hemlock was far more prevalent, preferring drier, more developed soils.[62] Some of these communities are still common today, although the massive exploitation of pines and hemlock in the late 19th and early 20th centuries changed these systems markedly, and they are still recovering toward a presettlement state.

The postglacial river valley tundra hosted musk oxen, mastodons, wooly mammoths, elk, caribou, white-tailed deer, wolves, giant beavers (larger than most black bears), and other large mammals. These species moved north as the ice retreated, and later many went extinct. The forests that replaced the tundra were home to smaller mammals. Early residents of the area's wetlands and lakes included loons, red-winged blackbirds, wood ducks, and muskrats living among the cattails, sedges, and other wetland plants. Eventually, beavers, otters, and white-tailed deer thrived in this wet, northernmost part of the valley.[63]

The Wisconsin River Valley has been home to a rich variety of people, beginning with Paleo hunters in pursuit of large game in early postglacial time. A recounting of Native American use of the river valley could start at the source of the river. *Lac Vieux Desert* derives from an Ojibwe word meaning "old planting grounds," referring to food gardens established by the Ojibwe at their summer settlement on the lake. Earlier, the headwaters region of the Wisconsin Valley was occupied by Woodland Dakota. Between 800 and 1100 CE, people built effigy mounds, remnants of which have been found at Lac Vieux Desert, Granite Heights north of Wausau, Plover, and Wisconsin Rapids. Farther downstream, near Wisconsin Dells, a late Woodland tribe, the Oneota, thought to be ancestors of the modern Ho-Chunk, lived in the valley between 1000 and 1600 CE, growing corn, beans, squash, and melons in raised gardens.

Over the centuries of Native American use of the river valley, the Wisconsin River served as part of an extensive trade network with trails connecting the Great Lakes to the Mississippi River and points west. Lac Vieux Desert was an important hub in this network, linking trails to the Wisconsin and Mississippi Rivers to the south; the Lac du Flambeau, Chippewa, and Mississippi Rivers to the west; the Montreal River and Lake Superior to the north; and the Wolf, Brule, and Menominee Rivers and Lake Michigan to the east. Within and among nations, Indigenous people traded many items, including beaver, bear, and deer pelts, for beads, necklaces, knives, and other tools and decorative items. Later, with European settlers, they traded for kettles, mirrors, firearms, and other conveniences.[64]

This network was also used by French explorers and fur traders, beginning in the 1600s. Among them were Father Jacques Marquette and trader Louis Joliet, who explored the river in 1673. Marquette was credited with first recording the river's name as *Meskousing*, an Algonquian word meaning "stream that flows through something red," likely referring to the reddish sandstone canyon of the Dells. The literature tracks this name as having morphed into *Ouisconsin* by 1718 and *Wisconsin* by the early 19th century. Note that the state is named after the river, not the other way around.[65]

Indian nations faced upheaval with the arrival of European explorers and fur traders. Some nations allied themselves with fur trading companies, others with competing colonial nations. This led to more aggression and conflict among some nations. With the growing influx of European settlers came diseases, including smallpox, cholera, and measles, for which the Indigenous people had no biological defenses. Over time, epidemics of these diseases wiped out entire villages along the Wisconsin River.

The Dakota lived peacefully in the valley of the Upper Wisconsin River until the late 1600s, when the Ojibwe began incursions into the valley from their home territory on the shores of Lake Superior. After two centuries of fighting, the Ojibwe had taken control of the Upper Wisconsin River Valley. Within another half century, by the 1850s, they in turn were forced to cede most of their lands to the US government.

The scope of this book generally does not include much detail beyond early human history, but for purposes of understanding the modern nature of the Wisconsin River, it is worth noting that the lumber boom of the late 19th and early 20th centuries changed the river dramatically. As loggers were cutting down the great pine forests, they used rivers to float large rafts of logs downstream to sawmills. On the Upper Wisconsin River, the many rapids became obstacles. Logging companies then built several dams, which created large pools for storing and moving the logs down river. Many of the original logging dams have been replaced by modern dams built for flood control and hydroelectric power. The Upper Wisconsin has 25 dams and 21 reservoirs, the largest of which are, in order, Petenwell Lake, Castle Rock Lake, and Lake Wisconsin.[66] Most of the rapids that once frothed along the length of the Upper Wisconsin River are now underwater behind such dams.

TRAVEL GUIDE
The Upper Wisconsin River

The Uppermost Stretch. Travelers can view the river's source, Lac Vieux Desert, in a public campground about one mile south of the Wisconsin–Michigan border on West Shore Road. Take this road south one mile to County Road E, then west for three miles to US Highway 45. Take 45 south 15 miles to the town of Eagle River. Here you'll find Watersmeet Lake—the first of several large impoundments, or reservoirs, on the Wisconsin River—where the waters of the Wisconsin River, Eagle River, and Mud Creek meet. Other popular major impoundments on the Upper Wisconsin include the Rainbow Flowage south of St. Germain, the Rhinelander Flowage just upstream of Rhinelander, and the Spirit River Flowage south of Tomahawk.

Grandfather Falls. South of Tomahawk on State Highway 107 is the Grandfather Falls Hydroelectric Plant's northern parking area, where you can hike about a quarter mile to see the cascade of rapids called Grandfather Falls. For a more scenic route, take 107 another two miles south to Camp New Wood County Park and hike upstream for about a mile to the lower end of Grandfather Falls on a section of the Ice Age National Scenic Trail. This riverside trail was tramped for centuries by Native Americans, early explorers, and fur traders, who used it as a portage trail around Grandfather Falls. These rapids drop a total of 89 feet in one mile, the sharpest drop on the Wisconsin River. The river flows over enormous Precambrian boulders, well rounded by centuries of flowing water and ice (Figure 2.24, left). They represent the roots of mountains heaved up between 1.9 and 1.8 billion years ago as a result of a collision between two primitive continents.

Council Grounds State Park. On the north side of Merrill, Council Grounds Drive runs a short distance from Highway 107 to Council Grounds State Park—a wooded park along the river where a dam created Lake Alexander. An open stand of stately white and red pines along a beach on this lake blends into a thicker forest of mixed hardwoods and hemlock, reminiscent of the forest types that existed here before they were exploited by logging companies and other industries. Native nations may have gathered here in the distant past for annual councils, ceremonies, and festivals.

Rib Mountain State Park. Southwest of Wausau, a quartzite mound called Rib Mountain sits 760 feet above the Wisconsin River, higher above its surrounding area than any other bedrock site in Wisconsin. From a 60-foot observation tower in the park, visitors view the Johnstown Moraine—the low north-south ridge that marks the western extent of the Green Bay Lobe. Between the park and the moraine are outwash plains where water flowed away from the most recent glacier as it melted over many centuries. On clear days, the view to the north includes the Wood Lake Moraine deposited by the Wisconsin Valley Lobe. Between the park and that moraine are prominent hills formed by much older glaciers and now drained by the Wisconsin and Rib Rivers, whose sprawling confluence is also visible from points on the park's trails.[67]

River Towns. The proud river towns of Wausau, Stevens Point, Wisconsin Rapids, and several smaller cities make use of the river for electric power, industry, and recreation. Each has at least one riverside park. The University of Wisconsin–Stevens Point Museum of Natural History informs visitors about the river valley's ancient and modern history.

Petenwell Lake and Petenwell Rock. South of Wisconsin Rapids, a sprawling flowage backs up behind a dam just northeast of Necedah. Petenwell Lake is Wisconsin's second-largest lake. At its south end where State Highway 21 crosses the Wisconsin River is a boat landing on the east side of the river. It commands a view of Petenwell Rock, the 100-foot-high butte of Cambrian sandstone on the west bank of the river (Figure 2.25). This is one of many sandstone buttes, spires, and mounds scattered across the sand plain—the ancient bed of Glacial Lake Wisconsin—and at one time it was probably an island in that lake. According to a local legend, two young lovers being pursued by disapproving elders fled to the top of the butte and jumped into the river, never to be seen again.

Buckhorn State Park. South of Petenwell Lake, the next major impoundment of the river is Castle Rock Lake. Buckhorn State Park lies on the south end of a peninsula projecting into the Castle Rock Flowage on the ancient floor of Glacial Lake Wisconsin. The peninsula is lined with narrow sloughs and marshes where floodplain forests grow, attracting diverse wildlife, including ducks, geese, herons, muskrats, beaver, otters, and mink. Uplands in the park are covered by a mix of prairie, oak savanna, and rare oak barrens—sparse, sandy grasslands scattered with oak trees. These ecosystems host sandhill cranes, wild turkeys, white-tailed deer, coyotes, black bears, mice, snakes, hawks, and owls. The park seeks to provide a wilderness setting for visitors, similar to what once existed in the Wisconsin River Valley.

Dells of the Wisconsin River. The city of Wisconsin Dells was named for the Dells of the Wisconsin River—a set of spectacular sandstone gorges carved

2.25 Petenwell Rock, on the south end of Petenwell Lake, is one of many Cambrian sandstone mounds.

in the blink of an eye (in geologic time) as most of the water from the vast Glacial Lake Wisconsin funneled through the location where an ice dam had been breached. The flood sculpted a canyon averaging 150 to 200 feet wide and up to hundreds of feet deep. Over time, the river has filled much of the gorge with sand and other sediments so that it is now as deep as 140 feet, with cliffs standing as high as 100 feet over the water. While most of the gorge has been preserved and looks much as it did when it was created, the river level is 16 feet higher than it was in 1909 when a dam was built in the city of Wisconsin Dells, putting many spectacular erosional features underwater.

The walls of the seven-mile-long gorge are Cambrian sandstone, deposited around 500 million years ago. Its many layers consist mostly of a weakly cemented and easily eroded rock type called Galesville sandstone, capped by a thin layer of a stronger type called Ironton sandstone. This erosion-resistant upper layer forms overhangs, cliffs, and other striking features and somewhat protects the softer stone beneath it. The sand grains in this rock were deposited by streams flowing to a Cambrian sea, and much of this sand was accumulated in dunes shaped by heavy winds over millions of years.

Evidence of the Dells' geologic history can easily be seen in the canyon walls (Figure 2.23). The layers of stone, both flat and cross-bedded, or tilted, reveal the work of flowing water and wind moving sand. In tributary canyons such as Witch's Gulch, visitors can see how the rushing meltwaters eroded the rock, spinning gravel and small boulders within whirlpools to carve large potholes. Exploring the river by canoe is possible, although tour boat and power boat traffic make it challenging. Commercial boat tours allow for good views of the Dells as well as a hike into Witch's Gulch, which is 100 feet wide at its mouth and narrows to as little as 4 feet wide between canyon walls that stand 30 to 40 feet high. A state natural area helps to protect the canyon walls and adjoining strips of land from development, and trails through the area provide limited access to the riverside.

Lake Country

If you could have flown over the Northern Highlands of northwest and north-central Wisconsin around 9,500 years ago, you would have seen a sprawling expanse of randomly arranged hills, ridges, and hollows—dirt-brown, rock-gray, and muck-black, with only scattered hints of green wherever lichens and primitive plant life had begun to take hold. While water flowing away from the retreating glacier would have been drenching the landscape for decades, there would have been little in the way of standing water. The story of the Northern Highlands is about how this drab landscape was transformed into a lush, green forested region graced by thousands of sparkling blue lakes (Figure 2.26).

One of the major features of the Northern Highlands is what geologists call hummocky terrain—random arrangements of hills, ridges, and hollows. The formation of this terrain began 2.5 million years ago when the first glacier pushed into the area, which was then probably a rolling, rocky plain. The ice crunched and dismantled much of this bedrock, picking up the pieces and moving them along while action within and under the ice mass ground boulders down into rock fragments and their constituent sand grains. Over tens of centuries, the glacier collected, pulverized, and moved rock in this way. Because the flow of glacial ice caused this material to migrate forward and up within the

2.26 Big McKenzie Lake near Spooner is a typical kettle lake in northwestern Wisconsin. NANCY SPOOLMAN

ice body, this process resulted in a mass of ice permeated, and largely covered, with sand, gravel, and boulders.

When the climate warmed and the most recent glacier began to deteriorate, the parts of the ice mass that were richer in glacial debris absorbed more solar energy than did the debris-free portions of the ice mass, and thus they melted faster. This differential melting created an icescape of hills, hollows, ridges, valleys, lakes, and streams atop the glacier. The flowing water atop the glacier carried much of the debris from high areas to low areas on the icescape. As the ice melted, those debris-rich low parts of the icescape became high points on the newly created landscape that emerged from beneath the retreating glacier. The relatively debris-free high points on the icescape melted away to become low points on the landscape. The effect is called topographic reversal.

During the Pleistocene epoch, this process repeated itself with each advance of another glacier, perhaps 15 or more times in all.[68] In the Northern Highlands, this repeating process gradually formed a broad zone of high-relief hummocky terrain with characteristic hills, hollows, moraines, outwash plains, drumlins, kames, and eventually, lakes.

Two glacial features especially important to the story of the lakes that define the watery landscape in Wisconsin's Northern Highlands are kettles and tunnel channels. The majority of the thousands of northern lakes are kettle lakes, formed in the depressions left by masses of ice that dropped off the receding glacier, were buried under sand and gravel, and took decades or centuries to melt away. Tunnel channels were created by some of those subglacial streams gushing out from under the retreating ice mass, sometimes under extremely high pressure. Dozens of these powerful streams carved roughly straight or slightly sinuous long and wide troughs across the land, lying perpendicular to the glacial margin.

From a bird's-eye view, shortly after the glacier was gone, the area was a broad region with hills of all shapes and sizes and variously shaped ridges of glacial debris. Rivers flowed through gaps in the ridges and beyond, across vast plains, and under it all, great masses of dirty ice were shrinking away ever so gradually. A century or two later, lakes appeared where those ice masses had been buried, scattered among the greening hummocks and across the plains. In Ontario, Canada, far to the north of Wisconsin, this process also formed what is now Kettle Lakes Provincial Park, which an Ontario Parks blogger described as "a land shaped by icebergs."[69] The same can be said of areas in northern Wisconsin.

Water flowed abundantly during those postglacial centuries, and winds blew fiercely in the warming region bordering the still frigid retreating glacier. Much of the landscape might have dried up and been blown away, but the region was plenty moist, having rainfall patterns similar to those of today. Plant communities evolved and fairly quickly secured the hilly landscape in the face of possible erosion. As the glacier retreated, this evolution followed the familiar pattern from tundra to forest—from lichens, grasses, stunted alder, and other pioneer shrubs to spruce forests and finally to the mixed hardwood and pine forests we see today.[70]

Of the kettles formed during postglacial centuries, those that lay below the water table were slowly filled by meltwater and groundwater springs. Depending on the volume of water flowing in, some were filled to become lakes and

2.27 Several kettle ponds in the Chippewa Moraine State Recreation Area might someday become wetlands.

others were filled just enough to become ponds, bogs, or wetlands (Figure 2.27). Those remaining above the water table stayed dry and became wooded hollows or grassland basins. Correspondingly, varying plants eventually populated the kettles. Aquatic plants settled into the lakes, and wetland species found the swampy kettles. In some kettle wetlands, vegetation became buried by debris to form peat, the partially decayed organically rich material that now serves as a fuel and as a garden fertilizer.

Peat bogs in Wisconsin number in the thousands. While some include ponds, others are filled with wetland vegetation, and some host wetland tree species, all of them formed within kettles. Unlike most kettle lakes in northern Wisconsin whose water continually seeps into the groundwater supply through sandy, gravelly lake bottoms (for which reason they are called seepage lakes), bog lake bottoms are sealed by peat and other organic matter. While seepage lakes drop a little in a typical summer due to increased seepage and use of water by plant life, bog lakes do not drop at all because there is no seepage. Eventually, most bog lakes fill in with vegetation and become wetlands.[71]

Kettle lakes seem to be arranged randomly, but in some cases, they lie in a pattern such as a line or an arc. These configurations can tell geologists about conditions on the glacial margin. Sometimes, a number of ice blocks were detached at once by a burst of floodwater, or a long section of ice would break away from the glacier along a deep crevice near the glacier's edge. These processes account for several chains of lakes in Wisconsin.[72] In a relationship similar to that involving topographic reversal and hummocky terrain, the shape of a kettle lake somewhat reflects the shape of the ice mass that formed its basin—from oversized snowballs to monstrous agglomerations. Kettle lakes come in all shapes and sizes—from small and roundish to vast and sprawling.

Many kettle lakes have no visible inlets or outlet streams to connect to one another but are nonetheless connected by an underground network of extremely slow-flowing water. Water seeping into such a network from the floor of one kettle lake will eventually flow through springs into other nearby lakes. In a sense, many lakes can be thought of as nodes within a vast body of water lying under and above the ground across a region once covered by glacial ice.[73]

Today, a bird's-eye view of the Northern Highlands on a sunny summer day would reveal a vast green expanse peppered by the blues of lake water and the darker greens of wetland vegetation. Some viewpoints would reveal that parts of the region possess more water-covered area than dry land. Many lakes are lined with human-built structures, but many are not, and from the air, parts of the region resemble what existed before European immigrants came to change much of Wisconsin's landscape.

TRAVEL GUIDE
Lake Country

The variety of northern Wisconsin's kettle lakes and bogs is astounding. Together, Vilas and Oneida Counties (Figure 2.1) have the highest concentration of lakes in the United States and one of the highest in the world. And kettle lakes are scattered across all or parts of 26 counties in northern Wisconsin. The Northern Highlands also have tunnel channel lakes and flowages created by dams. This guide provides examples of all three types of lakes.

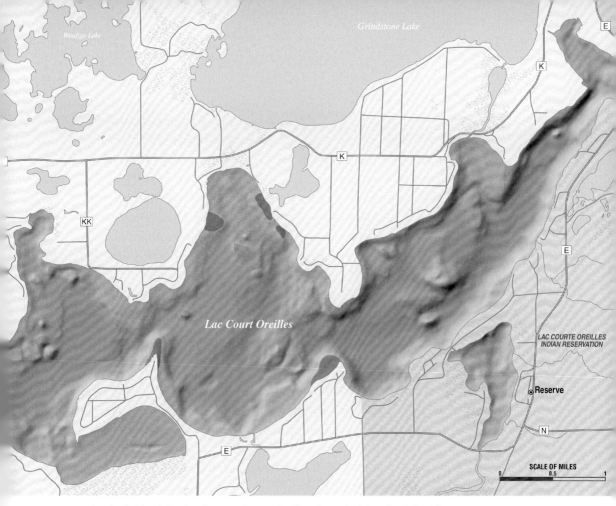

Windigo Lake

Grindstone Lake

K

K

KK

E

Lac Court Oreilles

LAC COURTE OREILLES
INDIAN RESERVATION

Reserve

N

E

SCALE OF MILES
0   0.5   1

E

2.28 The shaded land that borders Lac Courte Oreilles along the lake's far right side represents the Lac Courte Oreilles Indian Reservation. MAPPING SPECIALISTS, LTD., FITCHBURG, WI

Kettle Lakes

Lac Courte Oreilles. Located just east of State Highway 27 about eight miles south of Hayward, this sprawling 5,139-acre kettle lake (Figure 2.28) is one of a cluster often referred to as the Hayward Lakes. The eastern half of the lake lies within the Lac Courte Oreilles (LCO) Reservation where the LCO Band of Lake Superior Chippewa Indians reside. Native Americans have long valued the lake as a place to live and gather wild rice, as evidenced by signs of a village and mounds built more than 2,500 years ago on the south shore.[74] The LCO band has maintained a steady presence in the area to this day, harvesting fish and wild rice and establishing a K–12 school and a college on the reservation.

The Ojibwe, an Anishinaabe people, named the lake Odaawaa-zaaga'iganiing, meaning Ottawa Lake, in honor of the Odawa, another Anishinaabe tribe in the region. The ancestors of today's Odawa people followed a practice of altering the shapes of their ears, leading French fur trappers to call the local Odawa "Courte Oreilles," meaning "short ears." The lake was, in turn, named after those Odawa people.[75]

The lake is about six miles long and up to two miles wide with numerous bays and rocky peninsulas that give it an irregular shape. Typical of large kettle lakes, its basin features a number of high points, or sandbars, where the water is a few feet deep, and depressions, the deepest being 90 feet underwater. Partly because of this underwater structure, the lake is a popular fishing destination. It is connected by streams to its neighboring lakes, Grindstone Lake and Little Lac Courte Oreilles. The network of streams draining these lakes connects to the Mississippi River, thus connecting all these lakes to the Gulf of Mexico.

Chippewa Moraine State Recreation Area Lakes and Ponds. About 60 miles south of the Hayward Lakes area, at the southern edge of the Northern Highlands, lies a well-preserved section of the Chippewa Moraine, formed by the last glacier. The area is seven miles east of New Auburn on County Road M off US Highway 53. It features some of the best examples of hummocky, kettle-studded terrain in the United States. Geologists think the area was covered by about 150 feet of stranded, stagnant ice, which in turn was buried by glacial debris for several thousand years after the glacier retreated. The ice probably melted at a rate of a few inches per century and left a 10-mile-wide zone of high hummocks and kettles.[76] Trails from the David R. Obey Ice Age Interpretive Center wind around and among hummocks, wetlands, kettle lakes, and kettle ponds (Figure 2.27) in a deeply wooded setting. Hiking the trails can provide a sense of ancient history, for the area is much like it was before loggers and farmers changed the surrounding lands and waters dramatically.

Trout Lake. Vilas County lies in the heart of lake country, where a dense array of lakes has drawn people for thousands of years, from early peoples to today's lakefront landowners and thousands of tourists every year. Near the center of this array is Trout Lake (Figure 2.29), which is really two kettle basins joined by a narrows to form a 3,864-acre lake. Extraordinarily cold and deep, its northern basin reaches a depth of 90 feet and its southern basin 117 feet.

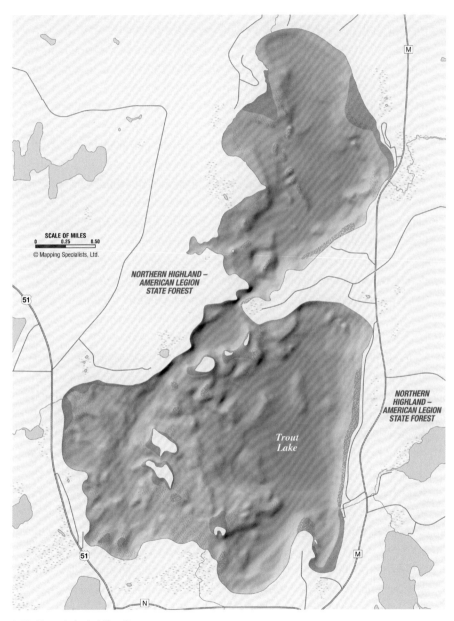

2.29 Trout Lake in Vilas County. MAPPING SPECIALISTS, LTD., FITCHBURG, WI

This area also attracts geologists and limnologists, or lake scientists. Geologists are interested partly because the lakes here happen to be located over the Niagara Fault, a 25-mile-wide zone stretching across northernmost Wisconsin. Trout Lake is somewhat central to this zone, which is bounded by lines running roughly east to west, a few miles north and south of the lake. The Niagara Fault zone is where the primitive Superior continent north of the zone collided with a volcanic island chain to the south around 1.9 billion years ago to begin the upheaval of the Penokean Mountains.[77]

The location of this area over the now inactive fault means that the bedrock under the Trout Lake area is a crazy quilt of Precambrian rock types including pink granite, pink rhyolite, black gabbro, and basalt, all created volcanically more than 1 billion years ago, not to mention gneiss and schist, some of the oldest rock in the state at around 2.5 billion years old. This unusual hodgepodge was created when the ancient rock types from two different continental masses were mashed together and churned for millions of years as mountains arose during the continental collision. The bedrock is covered with 90 to 150 feet of sand and gravel outwash from the glaciers. Water flows slowly but freely through this outwash, which slopes gently to the north and connects many of the area's lakes.

A good deal of the sand and gravel brought by the glaciers to this area lies in moraines left by the Wisconsin Valley Lobe of the glacier. The south basin of Trout Lake lies north of the Muskellunge Moraine, an east-west trending, discontinuous ridge of knobs of stony gravel. The north basin lies in a gap in Bowlder Moraine, a lower recessional moraine built by the retreating glacier as it halted for a time. Both moraines include kames, and around a dozen large drumlins lie just northwest of Trout Lake, oriented north-northeast to south-southwest—the direction of flow of the glacial ice. The drumlins are up to 1.2 miles long, a quarter mile wide, and 150 feet high.[78]

Lake scientists from all over the world have traveled to this area to work at the Trout Lake Research Station, a world-class site for limnology and a field station for the University of Wisconsin Center for Limnology since 1925. With a fully equipped lab and year-round housing facilities, the station serves a global network of scientists who study myriad freshwater issues such as invasive species, pollution, and climate change. The center also works to educate policy makers and the general public about such research.

Partly because 80 percent of Trout Lake's shoreline is within the Northern

Highland American Legion State Forest, some of the lake's ancient ecosystems have remained largely unchanged. Ample wildlife, including loons, other water-fowl, black bear, white-tailed deer, and several stressed songbird species such as Blackburnian, black-throated green, and yellow-rumped warblers thrive here. A state natural area southeast of the lake encloses a forest of white cedar, black spruce, tamarack, and balsam fir. Its floor, carpeted with sphagnum moss, hosts a rich diversity of bog plants including American starflower, Canada mayflower, bunchberry, and several orchids.

Tunnel Channel Lakes

Straight Lake State Park. Back on the northwest side of the Northern High-lands, Straight Lake in Polk County is actually a kettle lake but it lies in a tunnel channel, along with Straight River, which drains it. The state park was created to preserve one of Wisconsin's finest examples of a glacially formed landscape. Straight River has flowed in its linear valley since long before the last glacier finally shrank away to the northwest. Here, a subglacial stream carved a channel nearly 90 feet deep, up to half a mile wide, and 7.5 miles long. Over the centu-ries, the stream slowed and deposited sand, clay, and gravel, building eskers within the tunnel channel. Straight Lake State Park contains a segment of the Ice Age National Scenic Trail, which skirts the tunnel channel and Straight Lake. One trailhead is located on County Road I where it crosses the Straight River. The highway bridge provides a good view of the tunnel channel (Figure 2.30).[79]

Beaver Dam Lake. Many lake names apply to more than one lake. For example, there are 80 Bass Lakes in Wisconsin, along with 16 Little Bass Lakes and 42 Lost Lakes (all of which have been found by now). Of the three Beaver Dam Lakes, the one in Barron County (Figure 2.31) is unique because of the way it was formed. This long, narrow, and very deep lake lies in an old tunnel channel that crosses the margin of the farthest advance of the glacier's Lake Superior Lobe. That line, called the St. Croix Moraine, angles across the northwest quarter of this northwestern county.

The moraine marks the low ridge where the glacier sat for many centuries, beginning around 16,000 years ago, while a growing subglacial stream first trickled, then flowed, and finally gushed out from under the ice at the point where the city of Cumberland now sits. For centuries, this river spread untold volumes of sand and gravel in a broad, flat plain beyond the glacier's edge. At

2.30 The tunnel channel occupied by Straight River was carved by a subglacial stream thousands of years ago.

some point, the flow diminished and the river began to drop its load of sand and gravel within the tunnel channel it had carved. This happened in countless tunnel channels, but in this case, for some reason, ice began to flow back into the channel, partially blocking the tunnel exit. This caused the buildup of glacial debris in an unusually large and wide esker near the mouth of the tunnel.[80] At the same time, the partial blockage of the tunnel caused the already high water pressure within the tunnel to grow. With water volume and pressure building in the tunnel, the river gouged deep holes on either side of, and upstream of, the newly built esker.

Today, the result of this glacial activity is an eight-mile-long string of lakes averaging around a quarter mile wide. Sand Lake and Kidney Lake are located at

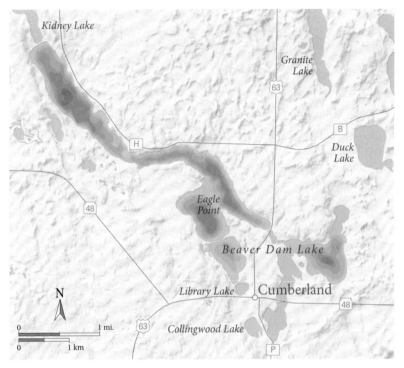

2.31 Beaver Dam Lake in Barron County. MAPPING SPECIALISTS, LTD., FITCHBURG, WI

the northwest end, and Beaver Dam Lake occupies the southeast half or more of the tunnel channel. The city of Cumberland sits on the large esker at the mouth of the tunnel and, because the lake more or less surrounds it, is known as the Island City. Northwest of town, where the glacier once sat, is classic hummocky terrain and lots of small kettle lakes. Southeast of town, the broad pancake-flat plain troweled by the subglacial stream is occupied by dozens of farms.

In Cumberland, the northwest end of the esker is the site of the popular Eagle Point Park, a hilly green expanse that commands views of the lake. On either side of Eagle Point, the lake drops off quickly to 90 feet, and at the northwest end of the lake, a little more than two miles northwest of the point, is a 106-foot-deep hole, which makes the lake one of the six deepest in Wisconsin. These three depressions are places where pressurized icy water must have formed ferocious whirlpools, drilling down into the deep glacial till over centuries as the water tried to escape the blocked tunnel under the melting ice mass. This amazing story is now commemorated on a colorful geologic marker in Eagle Point Park, created by the Ice Age Trail Alliance with guidance from UW–Madison geology professor David M. Mickelson.

Flowages

Hundreds of lakes in Wisconsin were created by the damming of rivers, and although they are not naturally occurring, they have played significant roles in modern cultures and economies. They also help to reveal parts of the geologic and natural histories of the region. I briefly explore a few of them here, starting with the Chippewa Flowage, created by the damming of the Chippewa River. This flowage, located southeast of Hayward, draws visitors by the thousands every year. It is Wisconsin's third-largest lake with 233 miles of shoreline and more than 200 islands (Figure 2.32). This is flooded hummocky land. The islands were once hills and ridges, the bays were all hollows and ravines, and the wildly irregular shape of the shoreline represents the random nature of hummocky terrain.

People go to the flowage to fish its waters and to experience the closest thing to wilderness in this part of the state. Being mostly undeveloped, the shores and islands remind visitors of vast tracts of Canadian wilderness farther to the north. The forest hosts aspen, birch, pine, northern hardwood, and oak— all species common to the forests that covered the state before loggers leveled most of them beginning in the 19th century. Flowage forests are home to many animal species, including bald eagles, ospreys, loons, great blue herons, several species of duck, white-tailed deer, black bear, wolves, otters, and other small mammals that were once far more numerous all over the state.

The creation of this flowage was controversial. In 1921, the state of Wisconsin authorized a Minnesota power company to flood the region over the strong objections of the LCO Band of the Lake Superior Chippewa. The flooding of more than 15,000 acres of land drastically changed the self-sufficient lifestyle of the Ojibwe, in part by eliminating 25,000 pounds of the annual wild rice crop. The Ojibwe had ceded the land to the US government in the early 1800s but retained the right to hunt, fish, and gather rice and other foods there. Some sites within the flowage area remain important for the religious, cultural, and economic purposes of the nation, which now operates the electrical power station at the dam and manages the flowage lands jointly with the US Forest Service and the WDNR.

The variety of Wisconsin's flowages reflect the varying landscapes surrounding them. For example, on the southern edge of the Northern Highlands,

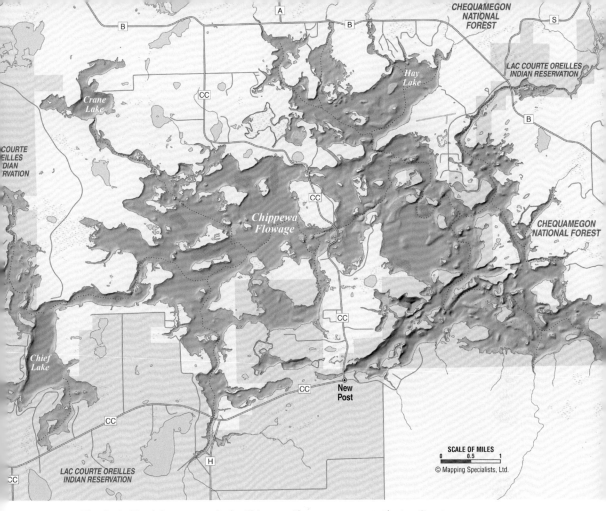

2.32 The shaded land that surrounds the Chippewa Flowage represents the Lac Courte Oreilles Indian Reservation. MAPPING SPECIALISTS, LTD., FITCHBURG, WI

the Holcomb Flowage is a long, shallow impoundment of the Chippewa River, lying on land shaped by old glaciers. On the north side of the highlands, in Iron County near Hurley, the Gile Flowage lies in an area where ancient bedrock is close to the surface, so its shores and islands are mostly bedrock. Another example of flooded hummocky terrain is the popular Turtle-Flambeau Flowage, a nearly 13,000-acre lake with 230 miles of shoreline and 195 islands. The thick birch and pine forests along the shores make it a haven for loons, eagles, ospreys, and countless other species. Much of this flowage is protected as a state natural area.

3.1 Northeastern Wisconsin. MAPPING SPECIALISTS, LTD., FITCHBURG, WI

N

| 0 | | 30 mi. |
| 0 | | 30 km |

VILAS

FLORENCE

ONEIDA

Menominee River

FOREST

MARINETTE

LINCOLN

LANGLADE

Lily

MENOMINEE

OCONTO

MARATHON

Upper Wisconsin River

Keshena

Green Bay

SHAWANO

DOOR

WOOD

WAUPACA

Wolf River

Green Bay

KEWAUNEE

PORTAGE

OUTAGAMIE

BROWN

Lake Michigan

WAUSHARA

MANITOWOC

ADAMS

WINNEBAGO

CALUMET

GREEN LAKE

MARQUETTE

FOND DU LAC

SHEBOYGAN

JUNEAU

3

The Northeastern Ridges
and Lowlands

The surface of northeastern Wisconsin was shaped by glaciers thousands of years ago, but its foundation was built by catastrophic events in Precambrian time. One of these was the continental collision that raised the Penokean Mountains 1.8 billion years ago, described in Chapter 1. Another was the formation of a distinctive oblong body of bedrock, called the Wolf River Batholith, lying under 3,600 square miles of land, roughly centered on the Wolf River Valley (Figure 3.1). Magma from deep in the mantle rose into the crust beginning around 1.5 billion years ago and forced its way into cracks and crevices in the granitic bedrock. It never erupted out of the crust which made it an anorogenic event, but its searing heat melted underground rock, creating metamorphic, granitelike rock types with names like monzonite, syenite, and anorthosite.[1] Some of these bodies of rock have a texture called *rapakivi*—Finnish for "rotten granite"—so called because it crumbles.

Such terms bring to mind the renowned geology writer John McPhee, who wrote about the lyrical nature of geological jargon. With tongue in cheek, he declared:

> Geology . . . was nothing if not descriptive. It was a fountain of metaphor. . . . Geologists could name things in a manner that sent shivers through the bones. They had roof pendants in their discordant batholiths, mosaic conglomerates in desert pavement. There was ultrabasic, deep-ocean, mottled green-and-black rock . . . festooned crossbeds and limestone sinks, pillow lavas and petrified trees, incised meanders and defeated streams.[2]

Surely, McPhee would agree that one description of the Wolf River Batholith, "a large anorogenic rapakivi massif," supports his argument.[3] Together, the Penokean episode and the Wolf River Batholith left a crazy quilt of various bedrock types now just under the surface of much of northeastern Wisconsin.

Much of the bedrock in northeastern Wisconsin was deposited by Silurian seas, which left multiple layers of dolomite and shale sloping gently eastward as they were deposited upon the Wisconsin Dome. Erosion, working from the top down, revealed not one underlying bedrock layer at a time, as it would have if the land were flat, but multiple layers resting side by side in long broad swaths. The softer rock layers, typically shale, eroded more quickly than harder dolomite layers—a process called differential erosion, which formed the alternating valleys and ridges that characterize northeastern Wisconsin. Because the broad dolomite ridges slope eastward, their eroded western edges often form steep sides called escarpments. The most prominent ridge in eastern Wisconsin is the Niagara Escarpment, which separates the lowland now occupied by Green Bay from that containing Lake Michigan.

Just before the dawn of the Ice Age, the Green Bay and Lake Michigan Lowlands did not contain large bodies of water; major rivers likely flowed northeast in both valleys, draining the surrounding highlands. During the Ice Age, several glaciers flattened the highlands and dug the lowlands a little lower, excavating softer sandstone and shale bedrock to form the broad basins of Green Bay and Lake Michigan.

The aftermath of the glaciers also played a major role in shaping northeastern Wisconsin. The Green Bay Lobe had blocked the flow of water draining northward in the Green Bay lowland. As the lobe began to retreat, its meltwaters backed up to form the vast Glacial Lake Oshkosh, which had multiple limbs that sprawled across the lowland. At its peak, around 19,500 years ago, it reached into southern Marinette County west of Crivitz. When the lake departed, it left a swath of sandy, rolling terrain through which many meandering rivers flow today. Together, these processes and effects tell the stories that follow.

THE MENOMINEE RIVER

The story of the Menominee River Valley is one of the oldest of the state's river valley stories. It starts with the collision of two continents—a cataclysmic event that built the Penokean Mountains between 1.89 and 1.82 billion years ago. It

3.2 Misicot Falls on the
Menominee River

created the unusual mosaic of bedrock that underlies much of northern Wisconsin and has strongly influenced the formation of the Menominee River's channel. As a result, the river flows through a cliff-lined, rocky valley for about 116 miles on the Wisconsin–Upper Michigan border and drains 4,100 square miles of land, a little more than a third of it in Wisconsin. Its course defines the knobby northeast border of Wisconsin (Figure 3.1).

The rocks underlying the Menominee Valley are even older than the Penokean Mountains. They are 2.3- to 3-billion-year-old metamorphosed granitic rocks (gneisses) that belonged to the embryonic core of North America, called the Superior continent, which was then north of what is now Wisconsin. A shallow sea lay to the south of the continent, and sometime between 2 billion and 1.5 billion years ago, due to atmospheric and ocean chemistry processes, large amounts of iron settled out in sediments in the quiet lagoons along the seashore. Geologic processes eventually converted these iron-bearing sediments into the reddish rock layers that are characteristic of the Lake Superior–area iron formations. These iron-bearing deposits played a major role in the history of the Menominee Valley.

The Penokean mountain-building, a 75-million-year process, actually involved two massive collisions, beginning 1.89 billion years ago. First, a chain of volcanic islands lying in a shallow sea south of the Superior continent were driven by tectonic forces to collide with the continent. The same tectonic forces later drove another small landmass from the south, called the Marshfield continent, into the islands, which by then were conjoined with the Superior continent. This infinitely slow but violent process churned bedrock and heaved up the Penokeans, a lofty chain that, by 1.82 billion years ago, stretched across northern Minnesota, Wisconsin, and Michigan and likely had about the same breadth and possibly the same height as California's Sierra Nevada of today.

The collision zone under these mountains is known as the Niagara Fault. Although it is inactive today, geologists can trace it from the Trout Lake area of Vilas County across the tip of Forest County to the Wisconsin–Michigan border, where it angles southeast across Florence County through the town of Niagara and on into Upper Michigan. This fault reveals the roots of the vanished Penokean mountains in a hodgepodge of bedrock types, including the ancient gneisses of the old Superior continent, formations of iron-bearing rock, metamorphosed sandstone (quartzite), and several others. All of these bedrocks were complexly folded and metamorphosed by the rock-melting heat and rock-crushing pressure of the mountain-building process.[4]

Those ancient rocks north of the fault include a striking formation called Quinnesec Volcanics, which can be seen along stretches of the Menominee River, especially the spectacular cliffs on the Michigan side across from the Wisconsin town of Niagara. These cliffs contain some of the oldest rock in the region—metamorphosed versions of black basalt from the ancient ocean crust; igneous rock such as gabbro, or "black granite"; volcanic rock such as the reddish rhyolite; thinly layered rocks called schists; and greenstones, or metamorphic rock stained green by certain minerals.

Other processes helped to shape the Menominee Valley, beginning with the raising of the Wisconsin Dome around 1.1 billion years ago (probably as a result of the Midcontinent Rift). This was followed by several advances of inland seas between a billion and 300 million years ago that left thick deposits of sandstone, dolomite (created by the remains of undersea plants and animals), and shale pressed from mud across most of Wisconsin. In the Menominee Valley, these rock layers slope eastward from the dome and down under Lake Michigan, so as the dome has eroded from the top down, beginning around 300 million years ago, the varying rock layers in the valley have become exposed. A key player in this erosion was the ancient Menominee River, which cut a channel across the entire sequence of tilted rock layers, from the Precambrian granites and schists exposed upstream to the younger sandstones and dolomites closer to the river's mouth.

Probably the biggest force of erosion shaping the Menominee Valley was Pleistocene glaciers. In the most recent glaciation, the Green Bay Lobe completely covered the Menominee Valley between 20,000 and 10,000 years ago. During its departure, the glacier dropped up to 400 feet of till in much of the valley region.[5] In the upper river valley (in today's Florence County), powerful meltwater streams deposited layers of sand and gravel onto a broad area of stagnant ice that had fallen away from the glacier. Over centuries, this ice melted out and the sediments collapsed to create hummocky terrain—rocky hills and low ridges interspersed with sandy plains, all loaded with water to be drained by rivulets, streams, and rivers or to sit in broad, poorly drained wetland areas.

As the Green Bay Lobe retreated, a vast meltwater lake called Glacial Lake Oshkosh (Figure 3.3) rose against the retreating ice wall and covered much of what is now the lower Menominee Valley along with Lake Winnebago and the Fox River Valley. Between 13,600 and 12,900 years ago, it covered an area of more than 2,500 square miles, draining south into the Wisconsin River near

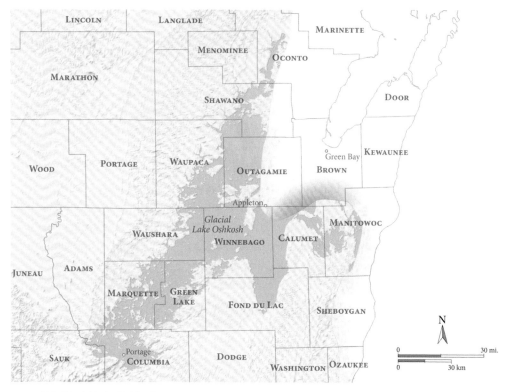

3.3 Glacial Lake Oshkosh. MAPPING SPECIALISTS, LTD., FITCHBURG, WI

Portage.[6] Over those 700 years, sand and clay flowed into the lake and were carried by currents to be spread in thick layers over large areas. For that reason, much of the lower Menominee Valley in Marinette County contains broad wetlands on clayey soil and is flatter and less hummocky than the upper valley.

This geologic story resulted in a picturesque valley channeling a river that was by turns ferocious and docile. Before extensive damming by humans, it flowed off the eastern edge of the Northern Highlands, raced through a hummocky, rocky upland, and then meandered lazily through a flat, wet lowland to its mouth on a major bay of Lake Michigan. Over its 116 miles, it dropped more than 700 feet.

Today's Menominee River begins at the confluence of the Brule River (which also forms a segment of the Wisconsin–Michigan border) and the Michigamee River, which flows south out of Upper Michigan. From there, the Menominee flows through forests of aspen, sugar maple, elm, yellow birch, and basswood;

limited stands of oak and pine; and wetland stands of spruce, balsam fir, and tamarack.[7] On its way to Green Bay, the river is fed by a number of large tributaries, including the Pine, Sturgeon, Pemebonwon, and Pike Rivers. Near its delta and mouth, now the site of the cities of Menominee, Michigan, and Marinette, Wisconsin, it crosses a nearly flat, sandy plain—the ancient floor of Glacial Lake Oshkosh. This 5- to 10-mile-wide plain stretches southwest and connects to the much broader Central Sand Plain. Southwest of Marinette, the narrow plain is dotted with ancient sand dunes, 10 to 30 feet high. Now largely hidden under sediments, the dunes were formed shortly after the glacial lake drained, exposing its sandy floor to the high winds of postglacial days.[8]

The postglacial watery environment of the lower Menominee became favorable to wetland plant communities that hosted wild rice. At the same time, lake sturgeon—a species that appeared on the planet around 136 million years ago when dinosaurs were still thriving—found the Menominee to be a perfect spawning grounds. The sturgeon is a goliath, sometimes growing up to seven feet long, weighing in at more than 300 pounds, and living as long as 150 years. The female, usually after reaching age 25, lays eggs every four to nine years. It is the oldest and largest native fish species in the Great Lakes. Sturgeon require the gravelly bottom and swift flow of a river to spawn, and then they need to return to big water such as that of Lake Michigan, which made the ancient Menominee River ideal for this species.[9]

For thousands of years, wild rice and lake sturgeon have been staple food items for people inhabiting the region. In particular, the Menominee, for whom the river is named, are the only Indian nation that originated within the borders of what is now Wisconsin. Menominee is an Algonquian term meaning "people of the wild rice."[10] When European explorers first arrived in the region in the 1600s, the Menominee Nation was the largest of the Angonquian-speaking nations in the region, which also included the Potawatomi, Sauk (or Sac), Meskwaki (or Fox), and Ojibwe Nations. According to French explorers of the time, a small Menominee village was located at the mouth of the river at least as early as the 1670s.[11] Oral tradition holds that the Menominee people's ancestral bear emerged from the Menominee River's mouth and was transformed by the Creator into the first tribal member.[12] (For more on the story of Native people in this area, see "The Wolf River" later in this chapter.)

Today's Menominee looks quite different than it did just after the glacier, mostly because people erected dams at several sites for hydroelectric power beginning in the early 20th century. A few rapids and waterfalls still exist, some of them quite dangerous, even for experienced paddlers, but many now lie underwater in the several reservoirs behind hydropower dams on the river. Waterfalls lost to damming include Sturgeon Falls, Twin Falls, and Little and Big Quinnesec Falls. Conservation measures have restricted shoreline development so that most of the reservoirs' shores are quite pristine.

The dams have had a major crippling effect on the entire sturgeon population of Lake Michigan. When the Menominee River was free-flowing, lake sturgeon would migrate by the thousands for more than 70 miles upstream from the river's mouth, seeking their age-old sites for spawning and rearing young. Fishery scientists estimate that the Menominee River alone provided this habitat for at least 45 percent of the Lake Michigan sturgeon population.[13] They calculate the lake sturgeon population before damming at around 2 million. It is now in the low thousands, due to damming, pollution, and overfishing.[14] State wildlife managers are attempting to remedy the situation by restoring some habitat and building passageways to allow sturgeon to reach their historic spawning and rearing habitats upstream of some major dams.

TRAVEL GUIDE
The Menominee River

Paddling the Menominee involves many portages around dams and some extremely dangerous rapids where a number of drownings have occurred. A safer option for most people is to tour the Menominee Valley by car. To get close to the river's source, go to Florence, Wisconsin, travel east on US Highway 2/141 for about a mile, and take Cross Cut Road north for a mile to Montgomery Lake Road. There, take a right and go east for less than a mile, and continue north on Camel's Clearing Road. Turn left on M-8 Road and follow it to a boat landing where the Brule and Michigamee Rivers meet to form the Menominee. For more downstream access points in this area, a good map can steer you down several other back roads that lead to the river.

Highway 2/141 continues southeast, roughly paralleling the river at a distance and crossing the river into Michigan just north of Iron Mountain. Just east of Iron Mountain, US Highway 141 departs from Highway 2 and crosses the river back into Wisconsin in the town of Niagara. Follow the highway south through town. It becomes Main Street and then River Street, on which you can park and

3.4 Cliffs of ancient volcanic rock across the Menominee River from Niagara, Wisconsin

get views of spectacular cliffs across the river on the Michigan side. On a clear day in late afternoon, the lowering sun lights the cliffs, revealing their multiple shades of reddish tan, brown, muddy green, and black (Figure 3.4). These are the Quinnesec Volcanics—that mix of rocks churned deep in the mantle during the Precambrian continental collision and then heaved up and sheared off by the mountain-building process. They represent the roots of ancient mountains.

A short drive past this viewing area on Highway 141, just east of Knutson Road, is a parking area for Sand Portage Falls, from which a half-mile hike takes you to views of a small waterfall and the upper end of Piers Gorge, carved into Precambrian rock by the river. Another short distance beyond this parking area, US Highway 8 joins 141 from the east. Take Highway 8 east and then north into Michigan to view the river's most spectacular rapids from the Michigan side (not accessible in Wisconsin).

Shortly beyond where Highway 8 crosses the river, take Piers Gorge Road to the west (left) for about a mile to the gorge. The road ends at a parking area where a foot trail takes you to views of the four "piers" of rock that form a series of short falls. These ledges of rock are the result of the faulting and fracturing that occurred during Penokean mountain-building. The highest ledge is under the eight-foot Misicot Falls (Figure 3.2). Here the river plunges around large boulders and into souse holes where boaters and rafters can get trapped under the roiling water. Kayaking or canoeing these falls is not recommended, even for the most experienced paddlers. The trail provides overlook views and closer looks at this waterfall and downstream rapids. The gorge is part of Michigan's Menominee River State Recreation Area.

Taking Highway 8 back south to where it joins Highway 141 in Wisconsin, you can continue south through the town of Pembine and another three miles to County Road Z, which runs east for 10 miles to the Menominee River State Recreation Area. Near the river, take Verheyen Lane north from Z a short distance to a parking area from which a set of loop trails traverses the woods, leading to an overlook view of Pemene Falls and to a canoe or walk-in campsite on the river.

Back on Highway 141, from the junction with Z, it is another 15 miles to Wausaukee. From here, follow State Highway 180 to the east as it roughly parallels the Menominee River all the way to the city of Marinette. (Along the highway between Niagara and Wausaukee are roads leading to the many spectacular and secluded waterfalls on other rivers that make Marinette County famous among waterfall enthusiasts; see next section.) A few miles east of Wausaukee, Highway 180 passes close to the river at an oxbow in the making. Here, the river has doubled back to meet itself, nearly creating a complete loop. Eventually, the river will shortcut through the top of the loop, north of the road, creating an oxbow lake. This is an indication of how wildly the river begins to meander as it crosses into a sandy, wetland-rich part of the state, entering the Green Bay Lowland. After Highway 180 veers south at McAllister, along with the river, it allows access to several views of, and boat landings on, the river.

The city of Marinette is named after Marinette Chevalier Jacobs, nicknamed Queen Marinette. She was the daughter of a Menominee woman and a French trapper, and she became an able fur trader. At around age 30, she entered into a common-law marriage with William Farnsworth, who operated a trading post at the mouth of the river in the 1820s.[15] This and other information on the area's history can be found at the Marinette County Historical Museum at 1650 Bridge Street in Marinette.

THE WATERFALLS OF MARINETTE COUNTY

Northeastern Wisconsin and Upper Michigan form a ragged funnel-shaped isthmus that arcs northeast between Lake Superior and Lake Michigan to the Canadian border. The wide southwest end of the isthmus is a rolling plateau lying about 1,000 feet above the two lakes and straddling the Wisconsin–Michigan border. On the southeast margin of that tableland lies Menominee County on the Michigan side of the border and Marinette County on the Wisconsin side (Figure 3.1).

Marinette County slopes off the edge of this low plateau, which exists because of the Wisconsin Dome, from which land slopes gently away on all sides but a little more steeply to the east. As noted in this chapter's introduction, a sequence of sloping layers of bedrock have been exposed by erosion working from the top layers down. Those upper layers are Paleozoic sedimentary bedrock, formed upon Cambrian sandstone, which was deposited on much older Precambrian crystalline rock, much of which was formed by Penokean mountain-building, described in Chapter 1.

3.5 Long Slide Falls on the North Branch of the Pemebonwon River

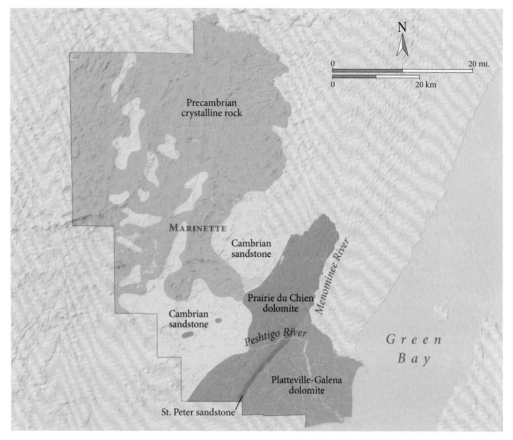

3.6 Marinette County spans a series of increasingly younger bedrock types (from northwest to southeast). MAPPING SPECIALISTS, LTD., FITCHBURG, WI

　　Marinette County contains this entire sequence as it dips from the highland gently down to the shore of Green Bay (Figure 3.6). The younger layers thicken toward the southeast, so successively younger rock types form the bedrock in this direction. The northwestern half of the county lies on Precambrian crystalline bedrock. Next in the sequence is a six-to-ten-mile-wide strip of younger Cambrian sandstone, at least 500 million years old. This is followed by a five-to-six-mile-wide strip of Prairie du Chien dolomite, a strip of slightly younger St. Peter sandstone that's less than a mile wide, and finally, a strip of Sinnippee Group dolomites that dive underwater to form bedrock under much of Green Bay. All of these later layers were deposited between 490 and 440 million years ago.[16]

In addition to Penokean mountain-building and the formation of, and erosion of, the Wisconsin Dome, two other geologic episodes added to the rich mosaic of bedrock in Marinette County. Both were intrusive events, when magma moved under the earth's surface without erupting from rifts or volcanoes. Instead, the magma forced its way into the rock matrix, prying open and filling crevices and caverns with searing hot fluid rock. These newcomer bodies of igneous rock became familiar types of granite now widely used to make countertops and other products. Around 1.86 billion years ago, a pink granite called Athelstane quartz monzonite intruded beneath a part of the county. Around 210 million years later came another intrusion of a gray granite called Amberg quartz monzonite. Both types are named for the Marinette County towns where they were first identified, and both have been actively quarried in the county.[17]

Surface features in the county reflect the differences in bedrock. Much of the northwest half of the county is hummocky—a patchwork of rocky hills and ridges, a few kettle lakes, and many low, swampy tracts. The southeastern half of the county contains more flat wetland and sandy plains. A countywide pattern of elongated lowlands and highlands reflects the sequence of bedrock—the softer, more erodible rock lying under broad lowlands. These lowlands channeled glacial ice again and again, most recently the vast Green Bay Lobe coursing south and southwest within the Green Bay lowland. The glacier built a series of north-south trending moraines in Marinette County as it receded. Thus, the major land features in the county—the gentle undulations from highland to lowland—are generally oriented northeast to southwest.

The lay of the land also determines the flow of Marinette County's waters and the shapes of its waterways. Because of the general southeasterly slope of the county's land as a whole, the drainage of its waters also trends southeasterly. The two major rivers, the Menominee and the Peshtigo, flow southeasterly (Figure 3.6), as do countless smaller streams. The Menominee and its tributaries drain 60 percent of the county's land, while the Peshtigo system drains the rest.[18] Many of these streams cross ridges of hard bedrock, where they have carved gorges and waterfalls.

Hence, Marinette County has an extraordinary number of waterfalls, especially on the Pike, Pemebonwon, and Thunder Rivers, earning the county the title "The Waterfall Capital of Wisconsin." The following travel guide describes a long tour of a sampling of Marinette County's dozens of waterfalls. Many of these falls are quite secluded and are accessible only via gravel roads. Even though they are not all entirely convenient, these destinations are remarkable for their scenic beauty and for the fascinating geologic stories they tell.

TRAVEL GUIDE
The Waterfalls of Marinette County

Smalley Falls. Morgan Park Road runs east from US Highways 8 and 141 about five miles south of the town of Niagara. Follow that road a short distance to the parking area for Smalley Falls. From there, it is a short hike to this waterfall on the North Branch Pemebonwon River. The terrain around the trail is somewhat rough, but this makes for a picturesque cascade of falls around small rocky islands. Here, the river has cut a deep notch in the ancient metavolcanic bedrock.

Long Slide Falls. Travel a little farther southeast on Morgan Park Road to the turnoff and parking area for Long Slide Falls. A quarter-mile hike takes you to this waterfall on the North Branch Pemebonwon River with options to view it from above or from the foot of the falls (Figure 3.5). Both views are stirring, and the upper view shows clearly how the river cascades 50 feet down through a fault zone in which layers of ancient rock were tilted to a nearly vertical position by the chaotic forces of mountain building.

Dave's Falls. Continue south on Highway 141 for 15 miles to just south of Amberg where Dave's Falls County Park sits on the west side of the road. This waterfall on the Pike River has two segments, the most dramatic of which is a narrow chute between two masses of granite (Figure 3.7). The park contains both pink (Athelstane) and gray (Amberg) quartz monzonite granitic rocks. It also features examples of smoothed rock polished by glacial ice, clearly visible in outcroppings between the parking area and the river. Somewhere near here in the days of the timber boom, a lumberjack named Dave was drowned trying to break a logjam on the river, and the falls were named in his honor. A footbridge takes hikers across the river to a more rustic trail along the riverbank. On all approaches to these falls, be careful. The trails can be rugged and the rocks slippery.

High Falls Dam and Reservoir. From Dave's Falls, drive another 20 miles down Highway 141 to Crivitz and take County Road A west and north about 10 miles to County Road X and turn west. After two miles, this road veers south to meet

3.7 Dave's Falls
on the Pike River

Deer Lake Road. Continue south on this road a few miles to High Falls Road, then turn right and go three miles to where High Falls Dam appears on your right. As you round the bend, the sight of the dam and falls can be arresting. This massive structure is 40 feet high, sitting on a 45-foot cliff in the broad valley of the Peshtigo River. Behind the dam, High Falls Reservoir sprawls across 3,600 acres and features large peninsulas and islands of Precambrian rock that resemble whales surfacing. These are the tops of high mounds of rock, now mostly underwater.

Veteran's Falls. Less than a mile downstream of High Falls, the Thunder River flows from the west into the Peshtigo River. Continue south from the dam on High Falls Road to Parkway Road, turn right (west), and go about half a mile to Veteran's Memorial County Park. At the south end of the park, steps take hikers down to the Thunder River, where they can hike on short trails to view three waterfalls, the uppermost of which is spanned by a footbridge. Veteran's Falls, while not spectacularly high or wide, are splendid examples of Marinette County's abundant outcroppings of rock more than 1.8 billion years old, exposed by centuries of erosion by flowing water.

Thunder Mountain Overlook. A few miles west of High Falls Reservoir, lying between the North and South Forks of the Thunder River and near their junction, is a 500-foot mound of quartzite known as Thunder Mountain. The name comes from a Potawatomi legend about thunderbirds that built their nests atop the mound, which the Potawatomi called *Chequah-Bikwaki*.[19] The mound is made of McCaslin quartzite, named after another, higher mound eight miles northwest of Thunder Mountain as the crow flies. These mounds are islands of metamorphosed sandstone and conglomerate standing above a rolling, hummocky plain that lies on the volcanic rock much more common to the area. The metamorphosis occurred during the formation of the Wolf River batholith some 1.5 billion years ago, and the resulting quartzite masses were later uplifted by geologic forces.[20] Thunder and McCaslin Mountains lie on the north edge of the batholith bedrock area.

To get to the overlook, continue west and north from Veteran's Falls on Parkway Road about two miles to Thunder Mountain Road and turn left. This road takes you 3.5 miles to a 160-acre park with a network of wide, well-maintained trails around the summit of Thunder Mountain. It is fairly heavily wooded, but the high points allow for limited views of High Falls Reservoir and the surrounding area. (McCaslin Mountain, straddling the Marinette–Forest county line to the northwest, is in a state natural area. While it is visible from some points on Parkway Road, it is more difficult to access than Thunder Mountain because it is not on public land.)

McClintock Rapids. For the next stop on this tour, go back east from Thunder Mountain to Parkway Road and continue north for several miles to County Road C. Along the way, Governor Thompson State Park offers a place to picnic and rest. Turn left (west) on C, travel half a mile and turn right (north) onto County Road I (a continuation of Parkway Road), and go about seven miles to McClintock County Park. Here, hikers can view modest rapids on the upper Peshtigo River from trails that lead through a shady hemlock forest, crossing four bridges over the rapids. The park is located in one of several scattered areas in Marinette County where glacial deposits on top of Precambrian rock are largely reddish-brown sandy till. This till is thought to have accumulated rapidly in massive movements of glacial debris, like mudslides or flooding rivers of meltwater. Such an event was likely caused by the sudden collapse of a mass of deteriorating ice around 10,000 years ago as the glacier was melting back.[21]

Strong Falls. From McClintock County Park, take Parkway Road (Highway I) about another two miles north to Goodman County Park Road, which forks to the left and, after another two miles, ends at Benson Lake Road in the park. Popular for camping and hiking, the park features Strong Falls (Figure 3.8), where the Peshtigo River crashes over volcanic ledges on both sides of an island in the river, accessible via a footbridge. The river cascades over numerous ledges and rapids—one of the most spectacular series of rapids on the Peshtigo River (Figure 1.5).

Eighteen, Twelve, and Eight Foot Falls. You can finish this tour by viewing three striking waterfalls on the North Branch Pike River. From Goodman County Park, take Benson Lake Road east nearly 10 miles to Old County A Road. Take this road north about four miles, turn right on Trout Haven Road and go east two miles or more, turn left onto Twelve Foot Falls Road, and drive north a short distance to reach Twelve Foot Falls County Park. The waterfall that gave the park its name is near the parking area, a frothy cascade that can be viewed from its top or

3.8 Strong Falls on the Peshtigo River

foot via hiking trails. Follow the trail downstream and through a campground to Eight Foot Falls, where the river has found a fast route between two large masses of Precambrian rock.

Back at the parking area, drive out of the park heading west, turn right on Twelve Foot Falls Road, and go north about a mile, watching for the sign to Eighteen Foot Falls (Figure 3.9). Take Eighteen Foot Falls Road east a few hundred yards to a parking area. The trail to the falls is a quarter mile long on rocky, quite rugged terrain. This is a favorite among waterfall hunters because of its remote, picturesque nature.[22] All of the falls at Twelve Foot Falls County Park tumble over granite forged by the Penokean mountain-building process more than 1.8 billion years ago.[23]

3.9 Eighteen Foot Falls on the North Branch of the Pike River

· ·

THE WOLF RIVER

The Wolf River is popular among canoeists and kayakers for its many exciting whitewater runs, all of which occur in the upper half of the river valley. There, the river dashes across dozens of ledges and boulder fields forming rapids and falls that are challenging even to the most experienced paddlers (Figure 3.10). The river drops more than 800 feet as it cascades through the upper valley, then slows abruptly, dropping less than 300 feet on the latter half of its journey. This dramatic change reflects the lay of the land between the river's source and its mouth and the complicated story of how that landscape was formed.

For purposes of this description, I divide the Wolf River Valley into three segments (Figure 3.1): (1) the northern segment, lying between the river's source in west central Forest County and the community of Lily at the intersection of State Highways 52 and 55 in Langlade County; (2) the central segment, between Lily and the town of Keshena in southeast Menominee County; and (3) the southern

3.10 Big Eddy Falls on the Wolf River

segment, between Keshena and the river's mouth at Lake Butte des Morts in Winnebago County. Each of these segments has its own geologic story.

The northern segment lies on the eastern margin of the Northern Highlands where the Precambrian granitic bedrock slopes gently southeast off the Wisconsin Dome and streams flow generally south and southeast. Several glaciers advanced across this area, the most recent being the Langlade Lobe of the Wisconsin glaciation, which arrived in the area around 22,000 years ago.[24] It flowed from the northeast to the southwest for many thousands of years, carrying untold tons of sand and gravel.

As the glacier advanced, it scraped away softer, wetter materials and sculpted long drumlins—teardrop-shaped hills—from the rock-hard frozen soils. As it melted, the Langlade Lobe draped the land with its load of reddish-brown sand and gravel in deposits up to 300 feet thick. The deeper deposits formed hummocks, or hills, and voluminous meltwaters spread the sand and gravel on plains lying among the numerous drumlins and hummocks. This outwash also buried stranded masses of glacial ice that eventually left kettles where lakes and wetlands would reside. Meltwater streams flowing through tunnels under the ice dropped sand and gravel that remained as eskers within the old tunnel channels after the ice was gone.

The postglacial upper Wolf River Valley was completely filled with glacial debris, and over centuries, the Wolf and its tributaries carved their valleys deeper, exposing the Precambrian bedrock that can still be found in several places. Over time, forests covered the higher ground. In low, wet areas, aquatic plants grew and decayed to form peat and muck atop the sandy, gravelly floodplains that are now broad wetlands lacing the hummocky terrain. Long forested drumlins stand prominently to the east and west of the Wolf River, which winds among the hummocks and plains of its uppermost reaches.

The central segment of the Wolf also flows within the eastern margin of the Northern Highlands and has a story similar to that of the northern segment, with some exceptions. One major difference springs from the formation of its bedrock, for this part of the Wolf flows from northwest to southeast over the Wolf River Batholith. Consequently, when erosion, particularly glaciation, dug the river valley deeper, it exposed types of granite that were different from those in the northern segment of the valley—younger types with different components that are more pinkish or reddish in hue, which formed under the surface where magma surged into the crust some 1.5 billion years ago.

Another factor that distinguishes the central segment of the Wolf River Valley from the northern segment is that, while the latter was glaciated by the Langlade Lobe of the most recent glacier, the central segment was molded largely by the gigantic Green Bay Lobe. Perhaps by coincidence, the western margin of the lobe moved roughly parallel to and close to the western edge of the Wolf River Batholith.[25] The lobe advanced across this area where previous erosion and glaciation had exposed some of the batholith bedrock, and it gouged out many boulders made of the pink and red granite. The till brought by the Green Bay Lobe contained pebbles of dolomite from bedrock to the east. The dolomite is yellowish-brown in contrast with the reddish-brown till of the Langlade Lobe, which contained more pebbles of basalt and iron-bearing rock from the north. Geologists use this till, among other forms of evidence, to determine how and where the two lobes flowed.[26]

The Green Bay Lobe flowed generally south, but on either side of its long axis, ice pushed laterally outward. In the Wolf River region, the ice actually flowed west and slightly northwest—upslope, because the land there slopes southeast. When the ice began to melt around 14,000 years ago, meltwaters flowed not away from the glacier, but southwesterly along its margin. As it receded from its farthest western margin, the glacier left a prominent north-south–oriented moraine, and meltwaters flowed between the moraine and the ice, creating a long, flat outwash plain of sand and gravel.

During the last centuries of the glacier, the climate oscillated between cooling and warming several times and the retreating glacier halted periodically, building a series of smaller moraines east of the terminal moraine, called recessional moraines. Each time the glacier began to melt again, waters would flow, forming an outwash plain behind the newest moraine. Today, the landscape in northeastern Wisconsin somewhat resembles a set of stair steps dropping gently in an east-southeast direction toward Green Bay. The tread of each step is an outwash plain, and the riser is a moraine. Several rivers today, including the Little Eau Claire, Plover, Little Wolf, and Embarrass Rivers, follow the routes of those old meltwater streams, parallel to the recessional moraines.

However, after the glacial retreat, some streams worked their way through the moraines and flowed southeasterly, as is the case with the Wolf River. On its upper reaches, it flowed parallel to the Langlade Lobe moraines. In east central Langlade County where the central segment starts, it began cutting across moraines of the Green Bay Lobe and descending those stair steps as it flowed southeast into the Green Bay Lowland. On this descent, the river also crossed hard rock ledges and boulder fields on the bedrock formed by the Wolf River Batholith, and this makes the central segment a roiling series of rapids and waterfalls (Figure 3.11).

South of Keshena, just north of the Menominee County line, begins the southern segment of the river—a different kind of stream altogether. Here the river has dropped off the Northern Highlands and into the Green Bay Lowland, where the land still slopes to the southeast, but only barely. The river, as if it has accomplished its mission, slows and becomes an easygoing wanderer.

The Green Bay Lowland was formed mostly by glaciers. They scooped softer stone out of the lowland, which became deeper with each passage of ice. As the Green Bay Lobe melted away, beginning around 20,000 years ago, the huge Glacial Lake Oshkosh (Figure 3.3) formed in the lowland. At various times during

3.11 Big Smokey Falls on the Wolf River

its 7,000-year lifespan, the lake encompassed all of the southern segment of the Wolf River Valley for centuries at a time. Where the river today crosses the Shawano–Outagamie County line, it passes through a 10-mile-long dune field. This is likely where the ancient Wolf and Embarrass Rivers flowed into the glacial lake, building a sandy delta.[27]

Over centuries, Glacial Lake Oshkosh collected tons of silt and clay delivered by rivers and wind, and these sediments sifted down to form a nearly flat lake bottom that slopes very slightly toward the southeast with the bedrock. The soil has high clay content, making it less permeable than other soils, and for these reasons, the Lower Wolf drops only 56 feet in 114 miles.[28] Its course—veering east, then west, crossing in and out of two counties and across two more—looks completely different from that of the Upper Wolf. An aerial view of the wandering Lower Wolf (Figure 3.12) presents a randomly careening stream with dozens of hairpin loops, s-curves, and turns of all shapes.

When the glacier finally retreated, it left this fascinating, dual-personality

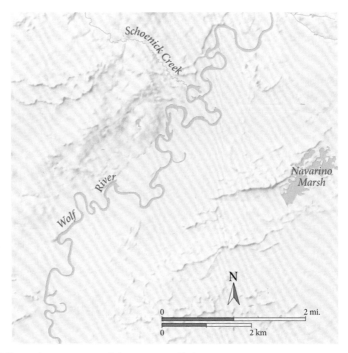

3.12 A wildly winding segment of the Lower Wolf River. MAPPING SPECIALISTS, LTD., FITCHBURG, WI

river that drains 3,690 square miles lying in parts of 11 counties, including almost all of Menominee County.[29] Its headwaters rise in west-central Forest County where a few small creeks feed Pine Lake, the source of the Wolf. From there, it flows 225 miles to its mouth in Lake Butte des Morts and the Winnebago Pool—a 265-square-mile region of waters, namely Lakes Winnebago, Butte des Morts, and Poygan and slow-flowing stretches of the Fox and Wolf Rivers.[30] Along the way, the Wolf picks up 20 tributaries, including the Pine, Rat, Waupaca, Little Wolf, Embarrass, Shioc, Evergreen, Lily, and Hunting Rivers.

The Wolf River Valley probably looks somewhat the same today as it did around 9,000 years ago after the last glacier departed. However, before the forests were logged extensively beginning in the late 1800s, they hosted much larger stands of pine and hemlock. In today's forests, these species are recovering along with spruce, balsam fir, and cedar. The valley lies across a transition zone between coniferous forests of the north and deciduous forests and grasslands to the south, so the lower Wolf River Valley tends more toward mixed hardwoods— maple, basswood, beech, aspen, and birch. Some 95 percent of the watershed is wooded and wild, hosting white-tailed deer, minks, foxes, otters, raccoons, black bears, and wolves. Birds living along the river include several duck species, herons, kingfishers, and eagles.

One animal that is far less common on the river than it once was is the lake sturgeon. For thousands of years after the glaciers retreated, sturgeon departed Winnebago Pool lakes in mid-April and traveled upstream 135 miles to Keshena Falls to spawn and rear their young. In 1872, the Northwest Improvement Company, working in the interest of the logging industry, began building a dam a few miles downstream of present-day Keshena near Shawano. This and other dams that followed blocked the sturgeon, denying them their historic spawning grounds, and their population plummeted. Historically, the Winnebago Pool was thought to have hosted the world's largest lake sturgeon population. It is now estimated to be a fraction of what it was before the dams were built.[31]

For many centuries, the sturgeon was a major food source for the Native people of northeastern Wisconsin. The dominant nation in the Wolf River Valley were the Menominee, whose name means "people of the wild rice" after the major food staple they harvested in the valley. Today, the river's name derives from the Menominee name, *Mahwaew-Sepew*—Algonquian for Wolf River.[32]

The story of the Menominee Nation is harrowing but also inspiring. When European explorers first arrived in the region in the 1600s, the nation numbered around 2,000—probably the largest of the Algonquian-speaking nations, which in Wisconsin also included the Potawatomi, Sauk (or Sac), Meskwaki (or Fox), and Ojibwe. In 1817, the Menominee occupied about 11 million acres in northeastern Wisconsin and part of Upper Michigan. That year, the US government recognized their territory in a peace treaty. In succeeding treaties, as part of a general government effort to get Native nations to define and then gradually cede their territories, the Menominee were cajoled and pressured into giving up their lands east of the Wolf River and north of the Fox River.

After the signing of an 1848 treaty, the US government further pressured the Menominee to move to Minnesota, but under the leadership of Chief Oshkosh, they refused to go. Four years later, the US military forced the Menominee from their ancestral land on Lake Poygan, near the mouth of the Wolf River, and onto reservation lands 70 miles to the north. In 1854, the Treaty of Keshena Falls established the Menominee Indian Reservation—235,000 acres in 10 townships straddling Shawano and Oconto Counties. This represented about 2 percent of the acreage the Menominee had inhabited prior to infringement on their land by the US government. Chief Oshkosh, whose name meant "the Brave," died in 1858.[33]

In succeeding decades, the Menominee endured many injustices at the hands of US government land managers and lumber and real estate interests. One such abuse was the clear-cutting of much of their remaining forestland in the early 1900s. During this period, the federal government enacted termination of several Native nations in an ill-conceived effort to free itself of certain obligations promised in return for the ceding of their lands. The Menominee's tribal status was officially terminated in 1961, at which time the reservation became Wisconsin's seventy-second county. This change allowed non-Indians to purchase and develop Menominee land. The nation's lands became subject to a complicated tangle of economic and political interests that led to a degraded economy and culture. After a long, trying 11 years, the Menominee convinced the government to restore their tribal status in 1972.[34]

This story includes other difficult chapters, which the Menominee survived partly because of their relationship with their beloved Wolf River and surrounding forests. Through Menominee Tribal Enterprises, the nation has managed its forests in such a way as to produce timber profitably while creating a model for sustainable forestry that has been studied by researchers from around the world.[35] Today, Menominee County is 95 percent forested, and the nation boasts 150 years of sustainable forestry. They are guided by the advice of Chief Oshkosh, who often repeated these words while northern Wisconsin forests were being cleared by the lumber industry in the early 1900s: "Start with the rising sun and work toward the setting sun, but take only the mature trees, the sick trees, and the trees that have fallen. When you reach the end of the reservation, turn and cut from the setting sun to the rising sun and the trees will last forever."[36]

Another hardship endured by the Menominee has been the disappearance of the sturgeon that once migrated annually to their waters. For centuries, the Menominee regarded the fish as their lifeblood—a replenishing gift that arrived every spring after the long winter. But that gift was denied them beginning in 1872 with construction of the Shawano dam and continuing through the 1920s, when more dams were built downstream of the nation's waters by loggers and electric power companies. Those dams blocked the annual sturgeon migration to their ancestral waters. In 1994, the nation began working with state and federal resource managers to restock the reservation's lakes with sturgeon from downstream in the Wolf River. In 2005, the sturgeon population in those lakes was large enough to allow for the first tribal harvest in more than 100 years.[37] Since then, restocking programs have continued and sturgeon have resumed spawning in their ancestral waters below Keshena Falls.[38] The Menominee Nation celebrates the revered fish in mid-April every year in the Sturgeon Feast and Celebration Powwow in Keshena, which is open to the public.

Several government protections help to keep the Wolf River relatively pristine. Its source lies in the 22,000-acre Headwaters Wilderness Area of the Chequamegon-Nicolet National Forest, where no development is allowed. Downstream in Langlade County, the Wisconsin Department of Natural Resources has designated a large area of state land around the river as the Upper Wolf River State Fishery Area, which greatly limits development on the riverway. Downstream, the 24-mile stretch with the most famous falls and rapids, all within Menominee County, has been designated one of Wisconsin's two National Wild and Scenic Rivers, which limits development in the river's watershed. The state has also designated the lower stretch of the river as the Lower Wolf River Bottomlands Natural Resources Area in an effort to sustain forests, wetlands, aquatic habitat, fisheries, and wildlife within floodplain forests, sedge meadows, tamarack bogs, and the other increasingly rare natural communities of the Wolf River Valley.

TRAVEL GUIDE
The Wolf River

Canoeing and kayaking are wonderful ways to experience the Wolf River, but it is a major challenge for even the most experienced paddlers. Here I describe several ways to see parts of the Wolf by driving and hiking. The river's source, Pine Lake, is northwest of Crandon off State Highway 32. A loop route connects to Highway 32 on both ends, and although most of the land around the lake is private, three public boat landings and a national forest campground allow access to this shallow kettle lake formed after the Langlade Lobe departed around 10,000 years ago.

Southeast of Pine Lake, sprawled over a vast area east of the Wolf River, is a spectacular field of drumlins molded by the Langlade Lobe. Some of these long, narrow hills stretch for more than a mile and rise more than 100 feet above the river floodplain to the west. The best view of these hills is on US Highway 8 between its bridge over the Wolf River and Crandon, where it rises up and over the huge drumlins. For a more rustic crossing of this drumlin field, take Old Highway 8 Road, which leaves Highway 8 just west of the river and veers east to cross the drumlins, connecting to County Road S just west of Crandon.

From Crandon, State Highway 55 runs south, rising and falling over more impressive drumlins and hummocks, before entering Langlade County. This drive provides more opportunities to see the work of the glacier in the upper Wolf River Valley. The road runs south to the community of Lily—a good place to begin a hike or bike ride heading south on the Wolf River State Trail, which lies close to the river in several spots between Lily and the community of White Lake in the southeast corner of Langlade County. (You can also take Highway 55 south and turn west on State Highway 64 to get to White Lake.) From White Lake, take County Road M south and then east to cross the Wolf River just north of the Menominee County line. At the crossing, just west of the river and south of the road, is a river access wayside with a large outcropping of Precambrian rock—distinctive pink granite, probably created by the Wolf River Batholith—lying at the river's edge.

The wayside is just across the river from the junction of County Road M and Highway 55, which runs south into Menominee County. There, many of the Wolf's most famous rapids can be found, including those of the picturesque Dalles Gorge (Figure 3.13) a quarter mile from the end of Dells Road; Big Smokey Falls, where the river drops 20 feet within 50 yards (Figure 3.11); Big Eddy Falls (Figure 3.10); and Keshena Falls—all of them special or sacred to the Menominee. All are accessible via hiking trails or dirt roads off Highway 55. In Menominee County, the Menominee Nation controls access to the river via a permitting process and runs a shuttle service for boaters. Visitors can get information and permits from the Menominee Nation headquarters in Keshena, where one can also visit the Menominee Cultural Museum to see exhibits on the history and traditions of the Menominee people.

Leaving Keshena, Highway 55 runs south for seven miles to Shawano, passing several river access points along the way. In Shawano, Sturgeon Park on the river is the last stop for migrating sturgeon in the spring. People travel here in late April and early May to witness the stirring sight of the sturgeons' spawning dance. The sturgeon circle each other and froth the waters as they shake violently in the process of laying and fertilizing eggs. Here, the Wolf has descended to the sand flats that, between 18,000 and 14,000 years ago, were the floor of the northern arm of Glacial Lake Oshkosh.

3.13 Lower Dalles of the Wolf River

Following Highway 55 south through Shawano and turning south onto County Road K brings you to the Navarino State Wildlife Area, the largest managed segment of the Lower Wolf River Bottomlands Natural Resources Area. Its 15,000 acres provide perhaps the best available sampling of the Lower Wolf River environs—a mix of sandy uplands and ridges with marshy depressions hosting diverse habitat types, including meadows, swamp conifer stands, bottomland hardwoods, pine plantations, and mixed hardwood forests. The Wolf River and a tributary, the West Branch Shioc River, run through it. Canoeing and fishing are popular on this meandering segment of the Wolf River.

GREEN BAY

About 3 million years ago, in what would become northeastern Wisconsin, a river flowed north in a shallow valley. Its bed was largely shale, and the stream was gradually scouring away this soft bedrock, carrying its particles downstream. The valley sloped gently to the east, so the river hugged that side of the valley where it ran against a harder type of stone, made largely of dolomite. It slid on by, leaving this long body of the more resistant stone intact and gradually carving its bed deeper. The stream collected tributary waters from the east and west and grew in volume as it flowed. The valley was probably mostly forested on the steep east side, and its west side sloped gently up onto a flat area containing broad wetlands and smaller forests stands.

Starting around 500,000 years later, the climate cooled and frigid winters lengthened, killing off the valley's forests and freezing its wetlands solid. The Pleistocene epoch had begun. The glaciers were coming. Massive walls of ice—maybe 15 of them over the next 2.5 million years—advanced from the north and gouged the valley deeper with each passage. When the climate warmed and the last of these great ice sheets had finally retreated to the north some 10,000 years ago, the now deeper valley was full of water—an immense bay on the west side of an even more colossal body of water that would become Lake Michigan (Figure 3.1).

The story of the valley where Green Bay formed really begins around 1 billion years ago with the formation of the Wisconsin Dome, which caused the Precambrian bedrock layers in the area to be tilted slightly down to the east. (Here "Green Bay" will refer to the bay itself as opposed to the city of Green

3.14 A view of Green Bay from Peninsula State Park

Bay.) Ordovician deposits in the Green Bay area were mostly shale, a set of rock layers called the Maquoketa Formation, ranging from 230 to 500 feet thick. It included a 50-foot-thick deposit of harder dolomite, sandwiched between deposits of shale, which now forms some of the lower cliffs along Green Bay's eastern shore.[39] Over the Maquoketa Formation, Silurian seas deposited thick layers of dolomite until about 417 million years ago. These layers, from bottom to top, are progressively richer in fossils, reflecting the growing diversity of marine life over time in Silurian seas. The dolomite layers now make up the long series of high cliffs that today border much of the east shore of northern Green Bay, known as the Niagara Escarpment.

The next stage in the Green Bay story was the long period following the withdrawal of the last ancient ocean more than 350 million years ago that left the Wisconsin Dome high and dry and subject to an eon of erosion. Geology professor John Luczaj of the University of Wisconsin–Green Bay has studied the area extensively. He wrote that this long period of erosion was "probably the most important interval of time in the geomorphic development of north-eastern Wisconsin." In the Green Bay area, he continued, "the predevelopment of deep-seated river valleys and some form of a precursor Niagara Escarpment must have formed during this time."[40]

We will never know what the area looked like at the end of that period, but geologists have made informed guesses based on what they know about the lay of the land. The eastern ridges and lowlands that characterize northeastern Wisconsin had begun to form, and in the Green Bay lowland, an ancient network of rivers flowed within an exposure of shale. The adjoining exposure of dolomite formed the early version of the Niagara Escarpment.

According to Luczaj, other deep-seated river valleys were likely formed during this period, one of them being the valley carved by the ancient Menominee River. It flowed southeast out of the highland as it does now, but then continued across the nascent Green Bay lowland, through a gap in the younger and much shorter Niagara Escarpment, and on into the valley of the ancient Michigan River, which flowed northeast through the next major lowland on the east side of the escarpment. The Michigan River Valley was the precursor of today's Lake Michigan Basin.[41] The gap in the escarpment is now called Sturgeon Bay. Any modern map of Wisconsin (Figure 3.1) shows that the lower Menominee River, as it flows into Green Bay, lines up nicely with the mouth of Sturgeon Bay on the Door Peninsula.

Farther north, from what is now the end of the Door Peninsula, the early Niagara Escarpment continued as a long, narrow dolomite ridge all the way to the Garden Peninsula, which now juts south from Michigan's Upper Peninsula. There were two major gaps in this stretch of the escarpment where ancient rivers had apparently been flowing off the ridge into the valleys on either side. The headwaters of these rivers eroded toward each other, weakening the escarpment at those two points and finally breaking it down to make notches. Erosion would eventually enlarge the notches to make the two big gaps, which were the precursors of the deep channels now called Death's Door Passage, between the Door Peninsula and Washington Island, and the Rock Island Passage on the northeast side of Rock Island.[42]

Another form of erosion at work during the long period since the ancient sea departed is one that continues today, namely the dissolution of dolomite by slightly acidic precipitation and groundwater. This process forms a type of topography called karst, which is characterized by crevices, caves, and sinkholes in dolomitic bedrock, and it has played an important role in creating rock features now found along Green Bay's shores and on the Door Peninsula.

The next act in the drama of Green Bay's creation was the Ice Age. The bay and its bordering escarpment would not exist but for the glaciers. Each one scooped another layer of slate from the lowland, but on the adjoining hard ridge of dolomite, they had a different effect. The Green Bay Lobe of the most recent glacier, like previous glaciers, overrode the escarpment, covering it with probably less than 500 feet of ice.[43] Under this ice, short periods of summer melting allowed water to seep into cracks and crevices. During the long glacial winters, this water refroze, expanded, and fractured the rock as the ice inched forward, carrying the fractured pieces with it. Thus, while the Green Bay Lobe bulldozed the lowland, it was more like a relentless wrecking crew up on the escarpment, steadily chipping away at that mass of rock. When it was over, the Green Bay basin had been carved to depths of up to 175 feet, while the escarpment remained high above the lake level along much of its length, reduced only slightly in its mass and height.[44]

As the Green Bay Lobe melted, water pooled in the lowland and lapped against the retreating wall of ice, forming Glacial Lake Oshkosh (Figure 3.3). Over centuries, the ice melted back and the lake found four major drainage routes into the Michigan basin, each north of the one that had formed before. Those routes are now occupied by, from south to north, the Manitowoc, Neshota, Kewaunee, and Ahnapee Rivers, the last of which flows across the south end of the Door Peninsula. These drainages are thought to have been catastrophic whenever an ice dam melted and released huge volumes of lake water, as indicated by boulders up to six feet in diameter found in some of the drainage channels today.[45]

The flow of meltwaters during the centuries of glacial retreat must have been a sight to behold. Before the ice had left the Sturgeon Bay area, a subglacial stream was gushing through the passage formed by the ancient Menominee River. These icy waters were carving Sturgeon Bay wider and deeper (Figure 3.15), pushing ice and glacial debris east into the Michigan Basin. As the ice retreated northward, slowly revealing the newly formed Green Bay, streams flowing off the escarpment and into the bay must have formed spectacular waterfalls.

The glaciers had carved the Lake Michigan basin hundreds of feet deeper than the Green Bay basin, so the latter was a hanging valley, separated from the deeper basin by a thinning segment of the Niagara Escarpment.[46] The meltwater must have flowed in torrents from the higher to the lower basin when the

3.15 Sturgeon Bay viewed from Potawatomi State Park

melting accelerated. Ice water and icebergs probably surged across the escarpment, tearing away rock and creating gaps where the mouth of the bay now lies, between Wisconsin's Door Peninsula and Michigan's Garden Peninsula.

Karst formation continued and probably accelerated during the time of prolific meltwater flowage. New cracks were etched into the dolomite surfaces on the east shore of the bay. Small cracks were widened into crevices and caves, and caverns were opened wider and made longer by gushing subterranean streams. This extensive network of karstic features enables water to flow freely down through the bedrock of the Door Peninsula. The last glacier left only a thin layer of sand and gravel atop the peninsula, so water now takes less than 24 hours to flow from the land surface down to the lake level in some areas of the peninsula.[47]

The period of glacial retreat was a time of slow-motion chaos for the Great Lakes, due to cycles of climatic warming and cooling, during which the level of ancient Lake Michigan rose and fell several times (see next section). In at least three instances, the lake's water level dropped so low that Green Bay emptied into the deeper Lake Michigan basin and became largely high and dry, except for self-contained pools in the deeper parts of the bay basin. When ice readvanced, the water level in the bay would rise again.[48]

All around Lake Michigan, including on the Green Bay shores, geologists have found wave-cut cliffs, terraces, and caves—all evidence of higher ancient lake levels (Figure 3.16).[49] Ancient beach lines visible above today's lake level present more evidence of higher lake levels, as well as of the rebounding of land compressed by the mass of glacial ice. On the north shore of Washington Island, one such beach line lies 95 feet above the present lake level. Tracing this old beach line south, geologists found that it slants down, lying only 40 feet above the lake on Sturgeon Bay. The northern stretches of this beach have rebounded 55 feet more than in the south, possibly because ice in the north was thicker and thus compressed the land much more.[50]

Today, Green Bay is a 120-mile-long, 10- to 20-mile-wide bay sprawling across parts of five Wisconsin counties and two counties in Michigan. From an astronaut's point of view, excepting the cities and other human artifacts along its shores, the bay looks much the same as it did a few centuries after the last glacier departed. At that time, the land sloped gently down to the water along most of the west shore. Three major rivers—ancestors of today's Menominee, Peshtigo, and Oconto Rivers—were building deltas into the bay made of the silt and sand deposited for centuries by a glacial lake and then washed out of the river valleys. And on the south end of the bay, waves and currents were building three east-west trending sandspits out of the glacial sediments on the bay floor. Today, these processes continue.

On the east side of the bay, the glaciers left the Niagara Escarpment standing as the backbone of the Door Peninsula. Along much of the southern part of the peninsula, the escarpment is a low wall paralleling the shoreline above narrow lowlands. On the upper half of the peninsula, the escarpment stands as a line of cliffs that get increasingly higher and hug the shoreline increasingly closer as one moves north. It is not continuous but rather is a series of bluffs standing as high as 150 feet above the water, punctuated by seven coastal lowlands where Door County's communities now lie.

3.16 This cave in Peninsula State Park was cut by waves of an older, higher version of Lake Michigan.

Beyond the northern tip of the peninsula in the mouth of Green Bay lies an archipelago, the remaining high points on the Niagara Escarpment segment that was ravaged by glaciers. Wisconsin's Washington and Rock Islands, Michigan's St. Martin, Poverty, and Summer Islands, and several islets among them are all remnants of the escarpment. Like the Door Peninsula, some of them have a steep west side on the bay, a gently east-sloping upper surface, and a low eastern shore on their Lake Michigan sides. Like the peninsula, they were carved and shaped first slowly by flowing water and then more ferociously by crushing masses of ice and their meltwater cataracts. Beyond those islands in the mouth of the bay is the northernmost segment of the east shore, Michigan's Garden Peninsula, also part of the escarpment. Its geology mirrors that of the Door Peninsula.

Of course, today's communities on the bay are not the first to exist there. People have inhabited the bay shores for at least 8,000 years, and the wealth of natural resources there have drawn people to the area throughout the millennia. The earliest inhabitants found ample forest plants, including berries, nuts, wild tubers, and greens, and they hunted deer, elk, beaver, bear, and raccoon. Over the years, Indigenous people increasingly fished the bay's waters for walleye, trout, whitefish, and lake sturgeon. Eventually, they used birch bark for making canoes and wiry shrubs to make fishing nets.

When European explorers ventured onto the bay in the 1600s, they likely found villages of the Menominee, Ojibwe, Odawa, Ho-Chunk, and Potawatomi along the shores. The first of these explorers was Jean Nicolet, who arrived in 1634, sent by the governor of New France in Quebec to scout the waters and forests in the interest of expanding the French fur trade. Reaching the shores of Green Bay, he formed a relationship with the Indigenous peoples living there.[51] The French established the first permanent European settlement on Green Bay in 1683.

From the later 18th to the early 19th century, the Menominee people became the predominant nation in the Green Bay area, having spread out from a homeland in the far northern Upper Peninsula of Michigan. As the French and British fur trades grew, the Menominee moved around the north shore of Lake Michigan into what would become northeastern Wisconsin, which they came to call their homeland.[52] They were the "people of the wild rice" and depended on sturgeon from the lake, wild rice beds on and near the shores, and the rich diversity of other resources afforded by living on or near the bay. Eventually, the US government reduced their territory from 11 million acres to today's 235,000 acres in Menominee County. (See "The Wolf River" section in this chapter for the story of the Menominee in northeastern Wisconsin.)

Much of the story of Green Bay can be traced by taking a driving trip around all or parts of the bay shore. Making the whole trip could take several days, but certain segments can easily be covered within a day or two. The following travel guide is divided into sections to serve both of these options.

Southwestern Shore

A good place to start a trip around Green Bay is in the city of Marinette, Wisconsin. The northwest shore and northern end of the bay are in Upper Michigan and could make an interesting side trip north from Marinette. However, this drive south from Marinette will give you a sense for how the Green Bay lowland dips gently east toward the bay, containing many shoreline wetlands, in contrast to the high rocky precipices that make up the east shore across the bay. This route crosses segments of a flat sandy strip of land that was part of the ancient floor of Glacial Lake Oshkosh, now lying just west of the bay, 20 to 80 feet above lake level.[53]

From Marinette, take County Road BB (also called Shore Drive) south along the shore on the delta of the Peshtigo River to where the road veers east to follow the river upstream. Just after that eastern turn, the obscurely marked Pond Road goes south off BB into the Green Bay West Shores State Wildlife Area, Peshtigo Harbor Unit. At the end of this short road, an observation platform allows a sweeping view of the broad, wet Peshtigo Delta, giving a sense of what the Green Bay shore must have looked like before European settlement. From here, BB runs west and northwest into the city of Peshtigo. A highly informative museum there commemorates the infamous 1871 Peshtigo Fire, which incinerated the city on the same night as the Chicago Fire, killing far more people than did that more well-known fire to the south. Both the Menominee and Peshtigo Rivers flow on ancient routes that have probably existed for millions of years, well before the glaciers arrived to excavate the Green Bay lowland.

From Peshtigo, take US Highway 41 west a short distance to County Road Y and go south for about 14 miles to Oconto. On the way, the North Bay Shore Recreation Area and D. E. Hall County Park provide access to the shore. On the west side of Oconto is Copper Culture State Park, which protects a burial ground containing remains of Old Copper culture Indians dating as far back as 5600 BCE. These people fashioned a variety of tools, weapons, and ornaments out of copper and are thought to be the first metalsmiths in North America.[54]

From Oconto, take County Road S south along the bay shore for 15 miles to County Highway J, which continues south paralleling the shore for 15 miles to the north side of the city of Green Bay. On the way, Oconto City Park and several units of the Green Bay West Shores State Wildlife Area allow access to the shore. In the city of Green Bay, many visitors like to get a look at Lambeau Field, home of the Packers, and visit Neville Public Museum, which informs visitors on the natural history of the bay area. The city's Bay Beach Park lies on the extreme southern end of the bay.

Southeastern Shore

From downtown Green Bay, it is 17 miles to Wequiock Falls, a popular destination for both amateur and professional geologists. Follow East Shore Drive to Nicolet Drive and continue northeast along the shore. Then, take a right on Van Laanen Road, turn left on Bay Settlement Road, and drive a short distance to the roadside park at the falls. A set of stone steps takes you down to the level of the lower falls for a close look. Here, you can see a full cross section of the Maquoketa Formation with the dolomite shelf sandwiched between thicker layers of shale, all more than 440 million years old. The upper set of falls is capped with Silurian dolomite that is much younger, only about 430 million years old.

Take Bay Settlement Road a short distance north and turn right onto Champion Road to access State Highway 57, which passes the Red Banks Alvar State Natural Area, where a parking area points hikers east, away from the edge of the escarpment. This site preserves a rare alvar plant community that grows on flat dolomitic bedrock with shallow soil. It hosts an unusual blend of boreal, southern, and prairie species, as well as one of the most diverse land snail communities in the Midwest. These communities are remnants of the postglacial environment, thousands of years old. The site also includes old stunted oak, red cedar, and common juniper.[55] This is a fragile and important member of Wisconsin's large collection of state natural areas.

From the state natural area, it is three miles north on Highway 57 to Bay Shore County Park. Here, on the steep road down the escarpment to the boat landing and beach, is an exposure of older Silurian dolomite. Near its base, this dolomite lies over the upper layer of the Ordovician Maquoketa Shale Formation, which is now covered by vegetation and talus, or rock debris, from the cliff above.[56] From the beach parking area, you can take a long metal staircase up to a trail that runs along the escarpment. There, you can view large blocks of dolomite that appear to be sliding down the slope (Figure 3.17). The dolomite blocks reveal the process of differential erosion, with the softer shale crumbling

3.17 Fractured, tilted layers of dolomite at Bay Shore County Park

beneath and the more massive dolomite cliff above fracturing and falling ever so slowly toward the bay.

The trip from Bay Shore County Park to Sturgeon Bay is 32 miles on Highway 57. Just west of the city, County Road PD (also called Park Drive) departs 57 going north to Potawatomi State Park. This park provides the first chance on this drive to see the high Silurian dolomite bluffs that make Door County famous. On the park's trails, hikers get panoramic views of Sturgeon Bay (Figure 3.15) from high up on the bluffs that make the south headland of the bay, as well as closer looks from wave-cut terraces along the bay shore.

Northeastern Shore

North of the city of Sturgeon Bay, on the north headland of the bay, is George K. Pinney County Park (also called Olde Stone Quarry Park), which you can access from Bay Shore Drive. This park offers striking views of Silurian dolomite in the walls of an abandoned quarry. You may notice distinct layers of stone, varying in texture and color. They represent 20 million years or more of ocean life remains that were deposited on an ancient sea floor.

The drive from Sturgeon Bay to Egg Harbor on County Roads B and G (Bay Shore Drive) is 17 miles, allowing some views of the bay. Between Egg Harbor and the end of the peninsula, the pattern of highlands alternating with lowlands becomes clear. It is 12 miles on Highway 42 from one lowland at Egg Harbor to the next at Fish Creek, the first entrance to Peninsula State Park. It is another two miles to Ephraim (the next lowland), nearly four miles to the next lowland at Sister Bay, six miles to Ellison Bay, and five miles to Gills Rock at the tip of the peninsula. Between each pair of lowlands, the highway rises up onto the escarpment, then descends to the next town. The pattern continues into the bay with the lowest areas under deep water and the highest points located on islands. Porte des Morts Passage (Death's Door) lies between the peninsula and Washington Island, where the high point is Boyer Bluff. On Rock Island, the highest bluff is called Pottawatomie Point.

These high bluffs all host a unique type of forest community—the cliffside white cedars—that laid claim to the bluffs in postglacial times. They are among the oldest trees in the Midwest, thought to grow as old as 1,200 years. These gnarled old cedars cling to small crevices in the rock and appear to be deformed as they grow downward or sideways from their fragile root systems. Some are almost literally hanging by a thread, kept alive by slender segments of phloem, the tiny pipelines that feed nutrients from the roots to the rest of the tree. Some that appear to be dead could have years or decades of life left in them. These old denizens of the escarpment are extremely vulnerable to even the smallest disturbances by hikers and climbers.[57]

Each of these bluffs of Silurian dolomite is a part of the Niagara Escarpment, lying on the west sides of the peninsulas and islands, and most of them are accessible via state and county parks. The most famous of these is Peninsula State Park, where gleaming white cliffs decorate a large peninsula jutting north into Green Bay from the Door Peninsula. The park's trails allow visitors stunning closeup views of such cliffs, including Eagle Bluff, where hikers get a dizzying view looking up at the towering precipice along with wave-cut caves and terraces formed by older, higher stages of the lake (Figure 3.16). Trails along the tops of some bluffs allow sweeping views of Green Bay and its islands and harbors (Figure 3.14).

Other access points on the peninsula and islands—Ellison Bluff County Park, Door Bluff Town Park near Gills Rock, Schoolhouse Beach Park on Washington Island, and Rock Island State Park—allow for additional good views of Green Bay and its fascinating bedrock features. Each has something unique to offer. Ellison Bluff provides a spectacular view from high over the bay and an observation platform built out over the edge of the bluff, allowing for a look straight down more than 100 feet—not for the faint of heart. Schoolhouse Beach is covered with cobbles the size of large potatoes. These perfectly smooth, oblong rocks were once bricklike pieces of nearby Boyers Bluff, made of a type of dolomite that fractures into such pieces. They fell into the bay and were tumbled and polished for centuries by lake waves that gradually moved them from Boyers Bluff to the beach inlet over thousands of years.

A trip to Washington and Rock Islands requires traveling by ferry, first from Northport, two miles east of Gills Rock, to Washington Island, and then from Jackson Harbor on the island's north side to Rock Island. It is well worth the time and cost. Tramping for a while on these ancient lake-bound fragments of the Niagara Escarpment is a wonderful way to wrap up an exploration of Wisconsin's Green Bay.

The Upper Lake Michigan Shore

Imagine turning the clock back to 10,600 years ago, going to the shore of Lake Michigan, settling into a beach chair, and then running the clock forward at one century per minute. At first, the shore would look much as it does today, but soon the waves would stop rolling in and the water on the shore would recede quickly out into the lake bed. It would be out of sight within 11 minutes, and after about 40 minutes, you would see the water line slowly returning. Within an hour, your beach chair would be under 20 feet of water and you would be best off swimming toward the new shoreline.

This is what happened, at a much slower rate of course, after the most recent glacier receded and the level of Lake Michigan fell and rose several times until

3.18 The upper Lake Michigan shore in Whitefish Dunes State Park

it found its present level around 2,000 years ago. The lake is now 307 miles long and 118 miles wide, with more than 1,400 miles of shoreline, 495 of which belong to Wisconsin. Its average depth is 279 feet, and its deepest point lies off the Door Peninsula under 920 feet of water. Lake Michigan is the second-largest of the Great Lakes, by volume, third-largest by surface area. It is really an inland sea—the only one of the Great Lakes that lies entirely within US boundaries. How did this enormous body of freshwater come to be here?

To begin that story, we need to turn our clock back much farther—about 545 million years to the beginning of Cambrian time and the Paleozoic Age. At that time, the area that would become the Michigan Basin was subsiding, or sinking, into a weak zone that had been created by the Midcontinent Rift half a billion years earlier. (The Michigan Basin is the vast area that today contains Lakes

Michigan and Huron with the Lower Peninsula of Michigan lying between them at the basin's center.) This subsidence, combined with the earlier formation of the Wisconsin Dome, caused the granitic basement rock of eastern Wisconsin to slope east off the Dome and down into the basin.[58]

Between 545 and 300 million years ago, a series of shallow Paleozoic seas invaded the continent and deposited layers of sandstone, shale, and dolomite onto the basement rock. Now, along Wisconsin's Lake Michigan shore, all of these layers slope east with an average tilt of half a degree, making for an easterly descent of a few inches per mile. This means the base of the Paleozoic rock layers lies 2,000 feet below the Door Peninsula and 16,000 feet below central Michigan—buried under succeeding rock layers and thick deposits of glacial till.[59]

The rock layers tell geologists that during the Silurian period, the middle part of the Paleozoic Age, the Door Peninsula shore of Lake Michigan lay in shallow ocean water on the northern edge of extensive coral reefs. The area was 10 to 20 degrees south of the equator, so tropical coral reefs and the variety of life they hosted thrived in warm, clear seawater, especially in the area now occupied by Milwaukee (see Chapter 4). Today's dolomite bedrock in eastern Wisconsin was formed from the remains of that sea life. On the upper shore of Lake Michigan, relatively dense concentrations of fossils can be found among the topmost (youngest) dolomite layers, indicating that sea life diversity increased here during the Paleozoic Age.[60]

After the last of the Paleozoic seas departed, erosion took over as the major geologic force at work in the Michigan lowland. Before the glaciers arrived, the lowland was probably drained by a river flowing northeast toward the area of present-day Lake Huron. The ancient Michigan River likely flowed near the center of today's Lake Michigan basin, eroding layers of shale and sandstone over many centuries to make its nascent valley deeper. (The Lake Michigan basin should not be confused with the Michigan Basin; the lake basin is a low part—a subbasin—of the much larger Michigan Basin.) Tributaries likely flowed to the lake basin from the west out of the rocky highland that would become the Door Peninsula. Flowing with the slope of the land, they found joints and cracks in the water-soluble dolomite and etched a series of shallow northwest-southeast–oriented valleys into the bedrock of the future peninsula.[61]

The many glaciers that inched across the land completed the work of the ancient Michigan River system, scooping deeper into the shale and sandstone with each passing. The last glacier's Lake Michigan Lobe took the lake basin's depth to more than 300 feet below sea level. The profile of Lake Michigan's floor is a telltale sign of such glacial scooping, based on geologic studies of modern glaciers around the world.[62] However, the glaciers were not able to do away with a strip of harder rock, called the Mid-Lake Reef, angling across the basin from the lakeshore between Sheboygan and Milwaukee to Muskegon, Michigan. Sitting nearly 200 feet above the lake bed's deepest point, the Mid-Lake Reef divides the basin in two, and the lake's water today swirls in two gyres—huge rotating currents—one in each part of the basin.

Like the Mid-Lake Reef, the dolomite in the highlands on Wisconsin's east shore—the young Niagara Escarpment and future Door Peninsula—was relatively resistant to glacial erosion as the ice masses plucked and pulverized smaller pieces of the escarpment but came nowhere near leveling it. Nevertheless, the ice carved the stream valleys of the ancient peninsula deeper, and a clear pattern began to develop there—bodies of hard dolomite separated quite regularly by these new stream valleys. This pattern now defines the Door Peninsula's topography.

When the glacier began to recede, this series of shallow valleys channeled growing streams of meltwater that continued to erode dolomite bedrock chemically, etching the valleys still wider and deeper. That meltwater became the new major force shaping the Lake Michigan shore. As the wall of ice retreated northward, the waters pooled to form a glacial lake that filled the newly shaped lake basin, lapping against that wall of ice. Meltwaters must have roared down through the peninsula's widening valleys, boring out large bays as they spilled into the lake.

Lake Michigan has had several lives, called stages, since the glacier started retreating out of its basin 14,000 years ago. In an early stage, when it was called Lake Chicago, water occupied the southernmost part of today's basin and covered what is now Chicago and northwestern Indiana. It drained south through the present-day Chicago River bed and across Illinois in the Des Plaines River bed to the ancient Mississippi River.

Sometime before 13,000 years ago, the melting ice uncovered a lower drainage route northeast into a glacial lake called the St. Lawrence Sea. Lake Chicago drained to a low level, occupying just the deeper central zone of today's basin. At the same time, north of the basin, meltwaters from the Lake Superior Lobe began to flow across Michigan's Upper Peninsula and into the Lake Michigan basin. This water carried red clay from Glacial Lake Duluth, where iron content in the Lake Superior basin had stained the lake sediments red, so this meltwater carried the reddish sediments into the Lake Michigan basin.

In the later centuries of glacial times, the climate wavered between cooling and warming, and the glacier readvanced twice into the Lake Michigan basin. The first advance occurred between 13,000 and 12,800 years ago. It covered nearly three quarters of the basin and reactivated the Lake Chicago drainage route. This ice mass picked up much of the red-stained sediments from Glacial Lake Duluth and, as it melted back, deposited them in a layer across the basins of Lake Michigan and Green Bay. Geologists have studied this and other sediment layers, using well-drilling reports and their own drilled samples to help assemble this story.[63]

By sometime before 12,000 years ago, the ice had again retreated, reopening the northeast drainage route. The lake again dwindled to a slender version of its present shape, and the Wisconsin shore extended for several miles into the basin. Over a period of three to four centuries, the climate warmed enough for forests to grow there. Evidence of these forests has been found near the village of Two Creeks at the border of Kewaunee and Manitowoc Counties. With additional similar discoveries over the years, geologists learned that this Two Creeks Forest—dominated by black spruce and tamarack—once occupied about 600,000 acres in the lakeshore region.[64]

Around 11,850 years ago, the climate again cooled and the ice readvanced to cover the northern half of the basin. Waters rose along the shore and drowned the Two Creeks Forest, which was then inundated by lake sediments, knocked down by advancing ice, and finally buried by the till left behind when the ice retreated.[65] This discovery site near Two Creeks, Wisconsin, now the Two Creeks Buried Forest State Natural Area, has played a large role in geological studies of the shore.

By 11,000 years ago, the ice had retreated for the last time. Lake Michigan's northeast outlet to Lake Huron through the ancient Mackinac River (today, the Straits of Mackinac) had connected Lakes Michigan and Huron to form Lake Algonquin, its level about 20 feet higher than those lakes are today. After a few centuries, the still-retreating glacier uncovered a lower outlet on the northeast corner of Lake Algonquin, at the site of North Bay, Ontario. It is likely that a dam made largely of ice had held back this vast lake and was finally breached, for the lake drained rapidly, its level dropping by more than 300 feet. Water exiting the Lake Michigan basin surged across the route of today's Straits of

Mackinac, gouging out a deep canyon that today lies under the straits and the famous Mackinac Bridge. The water eventually drained to the ancient St. Lawrence Sea, which occupied a vast area northeast of today's Lake Ontario.

Lake Algonquin's relatively short reign had ended, and a stage featuring the new smaller lake, called Lake Chippewa, had begun. By 10,300 years ago, this lake had drained to its lowest level, more than 100 feet lower than today's level, occupying only the deepest parts of the Lake Michigan basin—the two deeper zones separated by the Mid-Lake Reef. Again, Wisconsin's shoreline migrated to between 10 and 30 miles east of today's shore. Then, sometime after 9,500 years ago, the water level started rising again. This time, however, it was not due to a returning wall of ice.

Because Earth's crustal plates float on the fluid mantle, similar to barges on a river, they can be depressed into the mantle by a large enough mass. The North Bay outlet lay on land that had been deeply compressed by the most-recent massive ice sheet, and since that ice melted, the crust had been rebounding—and continues to rebound today. The segment of land under Wisconsin had been tilted down to the north, as the ice was more massive there. Hence, the rebounding of the crust raised the northern part of this segment and closed off the North Bay outlet of Lake Chippewa by raising it to a level higher than the present-day southeastern outlet for Lake Huron at Port Huron, Michigan. This change in the tilt of the land also caused the ancient St. Lawrence Sea to drain east, bringing Lake Ontario to its present level and creating the modern St. Lawrence River, which empties into the Atlantic Ocean.

Without the North Bay outlet, Lake Chippewa's level began to rise again, fed by rivers in the Great Lakes watershed that were in turn fed by meltwaters from the still-retreating glacier in the far north. This time, there was no wall of ice in the basin to contain the rising waters, and they gradually filled the huge basins of the present-day Lakes Huron, Michigan, and Superior, creating the largest of all lake stages, called the Nipissing Great Lakes, or simply Lake Nipissing. At its highest point around 5,500 years ago, this lake was about 25 feet higher than today's Lake Michigan. For centuries, as it gradually dropped to its present level, it deposited layers of sand and clay that today make up the beaches, clay bluffs, and other features on Wisconsin's lakeshore.

One such feature is the ridge-and-swale topography that has formed at certain locations where offshore waters are shallow and the land slopes up and away from the shore at a low angle. Lake Nipissing sat at its high level for centuries,

3.19 An aerial view of the Ridges Sanctuary in Door County, a prime example of a ridge-and-swale formation. © DOUG SHERMAN

its currents and waves carrying glacially deposited sand and building sandspits parallel to the shore. Eventually, a spit would grow to become a dune that would enclose the water behind it and form a lagoon. Meanwhile, the lake level was dropping due to the continuing retreat of the glacier far to the north. This would cause the lagoon's water to drain away, and vegetation would grow in the old lagoon, slowly forming a swale—a long, narrow low vegetated zone. At the same time, while grasses and other vegetation stabilized the dune to form a low ridge parallel to the swale, a new sandspit would be forming offshore.

Lake Nipissing's level continued to drop for around 2,000 years, but not steadily.[66] It seems to have stabilized and then dropped, again and again, at irregular intervals. Each time a ridge-and-swale pair would form over a period of decades, the lake level would drop again, and the process would repeat itself. Eventually, older inland dunes would become forested, while their accompanying swales would collect groundwater and precipitation to become ever-more diverse wetland zones. Today, areas with multiple ridge-and-swale pairs exist along the Lake Michigan shore (Figure 3.19).

Another prominent feature of the upper Lake Michigan shore is its regular pattern of bays and associated near-shore inland lakes. Each bay lies at the south or southeast end of one of the narrow lowlands running across the Door

Peninsula. Streams in these lowlands have carried copious loads of sand, silt, and clay down to the bays, making most of them quite shallow.[67] Among them, Europe, Kangaroo, and Whitefish Bays have been partially closed off from Lake Michigan by sandspits that grew to become dunes lying across the mouths of the bays. Over centuries, lake waves and longshore currents moved sand deposited by the glaciers to form these spits and dunes. The cut-off parts of these bays became shallow inland lakes—Europe, Kangaroo, and Clark Lakes, respectively.

Lake Michigan's sand dunes are remarkable examples of how natural communities can grow in places that seemingly could never support them. For example, it probably took 2,000 years for winds, lake waves, and currents to move and shape enough sand to build the dune that now encloses Europe Bay on the northern end of the Door Peninsula. At some point during this time, pioneer plants such as marram grass took hold in the deep sands, stabilizing the dune's surface to allow for more complex ground plants such as creeping juniper and, later, shrubs and trees to grow. Because of cool moist lakeshore conditions, dunes can host forests, including boreal forests—predominantly white spruce and balsam fir—that normally grow in more northern climes. Even wetlands can develop on old dunes where winds have scoured away enough sand to expose the water table. Here, groundwater seeps to the surface, creating a place where sedges and grasses can grow. Other plants soon find this moist area, and as they grow, die, and decay, soil is formed and a shallow marsh called a slack develops.

Evidence of the older, higher shores of Lakes Algonquin and Nipissing can be found on several parts of the Lake Michigan shore. There would be more such evidence were it not for bluff erosion that has shaped much of the Wisconsin shoreline, a process that continues today whenever lake levels are higher than normal. Most of the lakeside bluffs are made of clay and other sedimentary materials. High waves can erode the toe, or bottom section, of such a bluff, weakening the materials above and causing the face of the bluff to disintegrate and collapse under its own weight. It can take years for this weakening process to work its way up the face of a high bluff.[68] The collapse is usually sudden, which makes unstable bluffs dangerous places to hike or even to sit and enjoy the lake views.

The Lake Michigan shore's rich resources have drawn people to the area throughout the millennia. Paleo hunters arrived on the Door Peninsula at least 11,500 years ago. Centuries after the glacier retreated, their descendants found ample edible forest plants and hunted deer, elk, beaver, bear, and raccoons. Over the years, Indigenous peoples increasingly fished the lake's waters. The first people to settle more permanently were likely a group called the North Bay people, occupying the entire Door Peninsula between 100 BCE and 300 CE. Evidence of these people includes stone tools and pottery. They apparently traveled by canoe and used fishing camps from spring through summer.

Their descendants, following the Late Woodland tradition, were called the Heins Creek people, named for an archaeological site at the mouth of Heins Creek north of Whitefish Bay. They continued to develop fishing as a means of survival and established seasonal fishing villages. Their population is thought to have grown larger than that of any preceding group. During winter, they likely sought out the rock shelters and caves of the Door Peninsula. Some Late Woodland groups are known for their effigy mounds.

Descendants of this group, the Oneota, appeared on the Door Peninsula around 900 CE. They made improvements to earlier fishing methods and were also gardeners, growing corn and squash. With a greater variety of foods and better food storage, they may have been the first people to stay year-round in their camps, thus beginning the establishment of permanent villages. The Oneota may be the ancestors of the Ho-Chunk, who have had a major influence on the peninsula's history.

Other nations that have played major roles in the area are the Menominee, Ojibwe, and Odawa. The Potawatomi—whose name means "Keepers of the Fire"—were the dominant people on much of the peninsula when Europeans first began settling there.[69] The Lake Michigan shore and the Straits of Mackinac were important water routes used extensively by Indigenous people and by European explorers and traders. The word *Michigan* was the name given to the lake by the Ojibwe, derived from their word *michigami*, meaning "a great body of water."[70]

TRAVEL GUIDE
The Upper Lake Michigan Shore

Because of the eastward slope of the Door Peninsula's surface, the Lake Michigan shore has far more sandy beaches and low cliffs than the Green Bay side of the peninsula, with its soaring high cliffs. Wisconsin's Lake Michigan shore includes the eastern sides of Washington and Rock Islands (see the travel guide for Green Bay). This travel guide starts on the northeast corner of the Door

3.20 A terrace on Lynd Point Trail (left) and an ancient terrace cliff formed by Lake Nippising in Newport State Park (right)

Peninsula in Newport State Park, which features the sandy beaches and marshy areas so common along the shore, as well as low cliffs made of dolomitic rock, which can be seen along the Lynd Point Trail (Figure 3.20).[71] (This trail is accessible from Parking Area 3 in Newport State Park.)

These cliffs illustrate the lake waves' power in cutting terraces and shoreline cliffs. Waves of an earlier, higher postglacial version of Lake Michigan crashed upon the layers of rock overlying these terraces, breaking them apart and shaping the bench-like cliffs. Other abandoned shorelines can be found higher above the water and inland on trails such as the northernmost segment of the Lynd Point Trail (Figure 3.20). These ghost cliffs were carved by the waves and currents of Lake Nipissing beginning around 5,000 years ago.

For the next destination, take County Road NP from Newport State Park to State Highway 42. Turn left and go west and south to Sister Bay, and there, turn south on State Highway 57 for two miles to County Road Q. Turn left on Q to head east and south to Baileys Harbor, where the Ridges Sanctuary is accessible from the Cook-Albert Fuller Nature Center. Here, the lake created 30 ridge-and-swale pairs as it dropped in a stepwise fashion to its present level from that of Lake Nipissing more than 3,000 years ago (Figure 3.19). The sanctuary was established as Wisconsin's first land trust in 1937 and now protects more than 1,600 acres of shore land, including the ridges, which extend inland for a mile. They exhibit all stages of ridge-and-swale development, from open beach to dense boreal conifer forest, as well as more than 500 plant species, 60 bird species, 25 species of orchids, and 12 threatened or endangered species, including Hine's Emerald Dragonfly. Even a short hike in this pristine environment can provide a deep sense of peace and quiet.

The next stop, Cave Point County Park, offers spectacular views of more terrace cliffs and caves on the shore. Continue south on Highway 57, and just south of Jacksonport, turn left onto North Cave Point Drive. In two miles, turn left again on Schauer Road, which leads south into the park. Watching the waves crashing and spraying among the caverns and hollows under the cliffs will give you a sense of what it took for the flailing waves to sculpt caves here. The dense, hard rock that forms these cliffs is the uppermost, youngest layer of Silurian dolomite on the eastern end of the Niagara Escarpment, equivalent to the dolomite that caps Niagara Falls on the escarpment's western end. Called Engadine dolomite, it is light gray to purplish and can be identified by its deep horizontal creases and vertical crevices, chemically eroded over the millennia by precipitation and groundwater.

Cave Point is the northern headland of an old bay where Whitefish Dunes State Park is located, and its nature center is less than a mile farther down Schauer Road. (The county park is actually within the state park's borders.) This is the southernmost of the sites where waves and currents built a dune that closed off a deep inlet, this one called Whitefish Bay. The inland lake that was created is called Clark Lake (locally often referred to as Clark's Lake). The park occupies the wide, deep dune system that separates Clark Lake from Lake Michigan and the now much shallower Whitefish Bay.

At the north end of the park, stretching southwest from Cave Point, are low dolomite cliffs, rich in fossils from one of the planet's oldest coral reefs, which grew here on a Silurian seacoast more than 400 million years ago.[72] Fossil hunters look for brachiopods, fossil sponges (stromatoporoids), tubelike corals (*Syringopora*), chain corals (*Halysites*), and honeycomb corals (*Favosites*) (Figure 3.21). Fossil collecting is strictly prohibited, so take only photos and leave these important relics in place.

Along the shore and inland are trails through various dune ecosystems, including dune forests. On one set of boardwalks and staircases, you can climb to the top of one of the highest dunes on the Wisconsin shore, sitting more than 90 feet above the lake level. From there, you can view the lakeshore, as well as Clark Lake. If you could instantly turn the clock back 5,000 years, you

3.21 A *favosite*, or honeycomb coral, fossil on the shore in Cave Point County Park. This fossil is approximately 8 inches tall. GAIL MARTINELLI

would be perched high above icy waters at the head of ancient Whitefish Bay, which at that time stretched for miles to the northwest where Clark Lake is now.

Whitefish Dunes is also known as a site of important archaeological work. Much of what we know about Native American history on the Door Peninsula has been gained here. Archaeologists found evidence of a village site in the park that has been occupied by various groups of Indigenous people since at least 2,100 years ago. Near the park's nature center are reconstructed lodges and interpretive signs that inform visitors about early Native American life in fishing villages along the shore.

From Whitefish Dunes, take Nelson Lane or South Cave Point Drive along the shore to County Road T. Continue south along the shore on T for access to the Shivering Sands Unit of Cave Point-Clay Banks State Natural Area. It hosts classic lakeshore inland terrain with broad wetlands surrounding small undeveloped lakes that make excellent habitat for nesting waterfowl and wading birds. The wetlands are interspersed with stands of white cedar, balsam fir, and tamarack. The state natural area is laced with springs and streams that keep it saturated. It also encompasses a ridge-and-swale formation that supports a forest of white birch, red maple, beech, hemlock, and white pine—a community rarely found elsewhere on the Door Peninsula. Undeveloped hiking trails run into the area from two parking areas on Highway T southwest of Whitefish Point.[73]

At Lily Bay, Highway T veers inland. Shortly after leaving the shore, take North Lake Michigan Drive one mile south to County Road TT (Lake Forest Park Road) and continue south. Where this road passes a Coast Guard station, TT becomes Canal Road, which veers right and runs northwest along the Sturgeon Bay/Lake Michigan Ship Canal. Overlook Trail runs along the canal, accessible from several parking areas on Canal Road.

In Sturgeon Bay, where TT ends at State Highway 42/57, turn left and go south a few miles to County Road U. Then, continue south on U for about 10 miles to La Salle Park, built on high clay banks and named for French explorer Robert de La Salle, who is believed to have landed there in 1679.[74] From here, go south on U for three miles, then south on County Road S for a half mile, and turn right on Washington Road and go east for three miles to Blahnik Heritage Park on the Ahnapee River. This is a good spot for a picnic on a river that served as a drainage route for Glacial Lake Oshkosh as the glacier was retreating north more than 10,000 years ago. That meltwater stream engulfed the park area for centuries. The Ice Age National Scenic Trail runs along the river and through this park.

From Blahnik Heritage Park, it is another 25 miles on State Highway 42 through Algoma and Kewaunee to the Two Creeks Buried Forest State Natural Area. This is a small segment of a forest that grew here for a few centuries before being drowned by advancing glacial lake waters and buried by a readvance of the glacier thousands of years ago. Here on the eroding bluff face, sandwiched between two layers of glacial till, is a layer of dark brown to black organic matter. Within it, researchers have found well-preserved twigs, branches, needles and cones, and even tree stumps, some or all of which occasionally become visible in the steadily eroding bluff. This state natural area is also a unit of the Ice Age National Scientific Reserve. Collecting materials from the site is strictly prohibited.

The next stop is Point Beach State Forest on County Road O, which can be accessed by going south on Highway 42, east on County Road V, and then south on O. It features another series of forested ridges and swales formed off the prominent Rawley Point on the shore. State Forest literature describes 11 sets of ridges and swales, although some researchers say there are more.[75] The ridges vary in height from 3 to 26 feet high, and the swales are filled with peat deposits up to 5 feet deep. The most mature forested dune ridges host juniper, white pine, white cedar, hemlock, red oak, maple, and beech. Some of the swales stay wet, while others dry out by the end of summer. A set of trails provides a good sampling of the ecosystem that evolved here over 5,000 to 8,000 years.

Another good place to explore a pristine natural lakeshore environment is Woodland Dunes Nature Center and Preserve with its seven miles of hiking trails. Highway O continues south a few miles from Point Beach State Forest to rejoin Highway 42 in the city of Two Rivers. Take 42 to State Highway 310, or Hawthorne Avenue, and go west a mile or more to the site. It comprises more than 1,500 acres of mixed hardwood and conifer forests interspersed with wetlands and prairies, providing critical habitat for songbirds, bats, amphibians, monarch butterflies, and more than 400 plant species. With an elaborate boardwalk system, it is prized as an outdoor classroom for students of all ages.

The city of Manitowoc, five miles down Highway 42 from Two Rivers, is the end of the line for this travel guide. (The Lake Michigan shore travel guide is continued in Chapter 4.) Manitowoc offers the Wisconsin Maritime Museum, Red Arrow Park, and Silver Creek Park—all places to celebrate the natural beauty of the Lake Michigan shore, as well as its rich natural and cultural history.

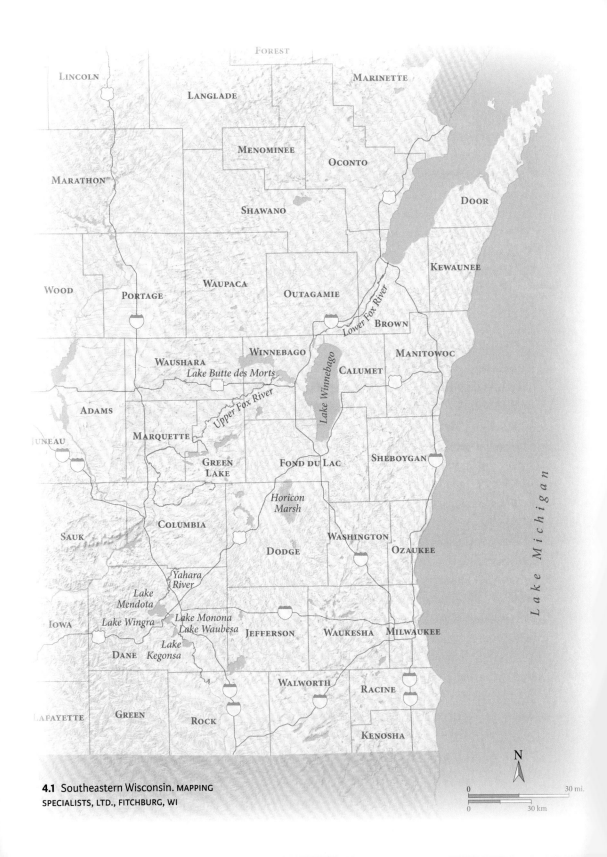

4.1 Southeastern Wisconsin. MAPPING
SPECIALISTS, LTD., FITCHBURG, WI

4

The Southeastern Glacial Showcase

As in northeastern Wisconsin, the bedrock in the southeast part of the state slopes to the east. However, in this case, it is not because of the Wisconsin Dome, but because of an extension of the dome called the Wisconsin Arch—a stretch of slightly elevated land running south from the dome toward the Baraboo Hills. Upon that sloping bedrock, the tropical Silurian and Devonian seas deepened gradually enough to allow for the formation of coral reefs—among the oldest on the planet (Figure 4.2).[1] These reefs and the ocean floor around them became deep masses of dolomite, which greatly influenced the formation of landscapes and waterways in the southeast.

Because of the eastward slope of the land, all Devonian and later deposits west of the Lake Michigan shore area were dismantled and carried away by erosion.[2] Only a narrow strip of Devonian dolomite remains along today's shore in Milwaukee and northward into Ozaukee County, and most of that is buried under glacial sediments.

Just before the Ice Age, southeastern Wisconsin probably resembled today's Driftless Area, with flat upland prairies, high ridges, and deep, forested valleys. The most recent glacier's Green Bay and Lake Michigan Lobes flattened and reshaped this part of Wisconsin. Along the shore, as the Green Bay Lobe retreated in fits and starts, it left a series of moraines lying parallel to today's lower shore, between 20 and 40 miles inland, called the Lake Border moraines.

Meltwaters from the retreating glacier also helped to shape the region's landscape. Those waters flowing northward into the Green Bay lowland backed up to form Glacial Lake Oshkosh, which, at its peak around 19,500 years ago, reached southwest as far as central Columbia County and southeast into Fond du Lac County. If the lake were to return suddenly today, the cities of Portage, Montello, Wautoma, Fond du Lac, and all of the Fox Cities would be underwater. Today, what remains of Glacial Lake Oshkosh includes Lake Winnebago and nearby smaller lakes.[3] However, the sandy, rolling lake bottom that developed over the 7,000 years of the lake's existence now forms the terrain through which many meandering rivers flow and on which eastern Wisconsin's large, shallow lakes lie.

The Lower Lake Michigan Shore

Around 420 million years ago, a primitive continent straddling Earth's equator was flooded by an inland sea. The sunbaked land around the shores of the sea was rocky and barren except for scattered, small, pale green plants and lichens. Under the water, however, was a world teeming with life. In the shallow areas of this sea, massive coral reefs had formed, hosting a colorful diversity of strangely shaped plants and animals in a scene that was utterly unearthly.

That continent was ancient North America, and the time was the Silurian period. The shallow sea covered most of the central part of the continent, lying between the present-day locations of New York and Nevada. Coral reefs—among the earliest on the planet—grew in the clear waters along several areas of this ocean's coastline. An especially rich collection of reefs grew along what would become the lower Lake Michigan shore in Wisconsin, centered roughly on the location of Milwaukee. Figure 4.2 depicts a diorama of such a reef, on display in the Milwaukee Public Museum.

The area around Wisconsin's lower Lake Michigan shore in Silurian time enjoyed conditions ideal for the growth of coral reefs—warm, shallow water kept clear of sediments by consistently strong currents. Coral reefs are built by tiny animals called polyps, closely related to jellyfish, that absorb calcium carbonate and secrete a limey substance to build a protective shell. When they die, their empty shells form part of a growing limestone platform for further polyp growth. Living in the tissues of the polyps are single-celled algae that, through photosynthesis, provide the polyps with nutrients and oxygen in exchange for their limestone shelters. It is the algae that give reefs their stunning coloration.[4]

In Silurian time, many creatures contributed to the reef structures, including delicately branching bryozoans, spongelike stromatoporoids, and at least 40 coral species. Over time, the porous rock platforms became complex structures

4.2 Diorama of a Silurian coral reef at Milwaukee Public Museum

with peaks, ridges, holes, and crevices that served as habitat for a growing variety of species. The following passage from the Milwaukee Public Museum's website rounds out the description of a Milwaukee reef:

> In this Silurian sea, stromatoporoids and tabulate corals built ancient reefs. Crinoid meadows and bryozoan thickets baffled the strong currents, while encrusting stromatoporoids and bryozoans bound and cemented the loose sediment and mud. Orthoconic nautiloid cephalopods jetted about the reefs searching for their prey—crawling, mud-burrowing trilobites—while dense patches of thousands of pentamerid brachiopods filtered the water for food.[5]

Reefs in the Milwaukee area are thought to have grown to 30 feet or higher above the sea floor, while reefs in deeper waters in the Chicago area may have

grown to more than 300 feet high. Scientists think they were the largest biological structures on Earth at the time, hosting the planet's most richly diverse
natural communities.[6] Between the numerous reef mounds in Silurian seas,
the remains of myriad sea creatures collected to form layers of dolomite that
encircled and engulfed many of the reefs. Geologists refer to the type of stone
in these reefs and in surrounding sedimentary rock layers of the same period as
the Racine Formation, named for its exposures in quarries in Racine, Wisconsin.
It includes an exceptionally pure form of white to light gray dolomite, which is
often rich in fossils.

The abundant reefs of the Silurian "age of corals" were eventually buried
under sediments of later invading seas and by other geologic processes. The
Devonian period followed the Silurian, and much if not all of Wisconsin was
covered by the Devonian sea. This ocean also hosted diverse life forms, but the
mix differed from that of the Silurian. Gone were the corals, crinoids, and ceph-
alopods, but added to the scene were more species of shelled animals and several kinds of primitive fishes. Some fishes had the ability to breathe air part of
the time and may have been forerunners of various land dwellers. They might
have gradually moved from the abundantly swampy shore areas onto dry land
covered by increasingly lush jungles that were evolving in this still-tropical
environment.[7]

All Devonian and later dolomite deposits west of the Lake Michigan shore
are now gone, eroded away. Only a narrow strip of Devonian dolomite remains
along today's Lake Michigan shore in Milwaukee and northward into Ozaukee
County, and most of that is buried under glacial sediments. Most of Wisconsin's
lower Lake Michigan shore area lies atop layers of Silurian bedrock, which add
up to 600 feet thick in places. Perhaps it is fortunate that erosion removed the
upper layers, for had it not, we might never have known about the underlying
masses created by the creatures that flourished on ancient reefs. These scattered
mounds make up parts of the bedrock in the shore area.[8]

After the last Paleozoic sea departed, life continued to evolve rapidly as the
continent was pushed by plate tectonics on a curving northward path toward
its present-day location. Before the glaciers arrived, the Lake Michigan shore
area of today lay in an ancient valley, probably forested and occupied by plen-
tiful animal life. The valley that would become the Lake Michigan basin was
being carved by a river flowing northeast. In the present-day lower shore area,
the Michigan River's tributaries drained the sloping land to the west, etching

the courses of some of today's rivers, including the Manitowoc and Sheboygan Rivers, as well as upper stretches of the Milwaukee River.[9]

The Lake Michigan Lobe of the last glacier covered the entire lake basin and strips of land 20 to 40 miles wide on the present-day lake's east and west shores.[10] It occupied all or parts of the basin for thousands of years, fluctuating slightly in size as the ice shrank back and readvanced several times. Each time the ice retreated, meltwaters would form a glacial lake in part of the basin. Today's shore area of the lake was molded primarily by the two most recent advances and their associated glacial lakes, between 13,000 and 10,000 years ago.

Some 10,300 years ago, as the glacier retreated for the last time, Lake Michigan dwindled to a much smaller version of itself. This shrinkage revealed a reef complex that divided the basin in two, with the south basin lying offshore of present-day Milwaukee. There, a smaller lake lay several miles out from today's shore. It was about a quarter of the size of today's lake, connected to a larger lake in the north basin by a stream flowing through a gap in the dividing Mid-Lake Reef complex. These two lakes together were called the Lake Chippewa stage of Lake Michigan's history. (For more details on the shaping of the lake basin, see "The Upper Lake Michigan Shore" in Chapter 3.)

The glacier left the Lake Border moraines alternating with lowlands carved by meltwater streams. In those valleys, streams such as the Milwaukee River still flow today.[11] Since glacial times, erosion has whittled those moraines down, leaving mostly hummocky higher land in their place.

Much of the lower Lake Michigan shore is lined with bluffs ranging from 20 feet high in Manitowoc County to 120 feet high in Ozaukee County. They are made largely of glacial till, held together by clay and anchored by plant roots. This makes them vulnerable to erosion by wave action, heavy precipitation, and disturbance of their plant cover. When lake levels are high, wave action can weaken the toe of a bluff, which destabilizes its face and can eventually cause it to collapse, usually suddenly and without warning. This makes unstable bluffs dangerous places to visit.

Another feature of the shore is the sand dunes that grace several locations, including the two state parks on the lower shore (Figure 4.3). They began to form as lake levels dropped, beginning 5,000 years ago after the historic high level of the Nipissing Great Lakes. As more and more lake-bottom sand was exposed, waves brought it to shore, where winds blew it farther inland. Grain by grain, winds built sand dunes over hundreds of years—the blink of an eye in

4.3 Dunes are the central feature at Kohler-Andrae State Park.

geologic time. As the dunes grew, they behaved like slow-motion water waves, moving inland on the shore, to be replaced by newly forming dunes closer to the water. Older dunes migrating inland were eventually stabilized by pioneer plants such as marram grass, which allowed for more complex ground plants like creeping juniper to take hold. Over time, shrubs and trees grew, and even wetlands formed on old, stabilized dunes.

The huge volume of water in Lake Michigan has created a cool, moist shoreland climate regularly affected by fog, spray, and storm waves. Winters tend to be slightly warmer and summers slightly cooler on the shore than they are farther inland, and annual precipitation tends to be higher. This makes for a unique mosaic of natural communities, including stands of eastern hemlock near the shore; forests of sugar maple, American basswood, and American beech farther inland; and dune communities that include white and red pine and rare or threatened species such as dune thistle. Since European settlement in the 1800s, many of these communities have been disrupted or replaced by farmland.

So favorable was the Lake Michigan shore's climate, and so rich its resources, that it has drawn people to the area for thousands of years. The nations that dominated this area before the time of European exploration and settlement were the Ho-Chunk and Menominee. Later, the Potawatomi played a major role, largely displacing the other groups; they were eventually displaced by European settlers and the US government. (For a more detailed account of Native American history along the Lake Michigan shore, see "The Upper Lake Michigan Shore" in Chapter 3.)

TRAVEL GUIDE
The Lower Lake Michigan Shore

The list of sites to visit on the lower shore begins with Fischer Creek Conservation Area, a Manitowoc County park on County Road LS with a hiking trail along the shore. North of Fisher Creek, the trail lies above the bluff. It descends to the beach where the creek empties into Lake Michigan, crosses the creek, and then ascends to the bluff view again, south of the creek.

The historic lakeshore city of Sheboygan is 11 miles south on LS. Here, in 1961, geologists drilled through bedrock and obtained a core sample containing the entire Paleozoic sequence of rock layers, from the Silurian through Ordovician and Cambrian right down to the Precambrian basement rock, 1,875 feet below the surface.[12] Such drilling is one way that geologists have mapped layers of bedrock, and this sample helped show the degree at which the bedrock slopes eastward at the shore. Sheboygan boasts expansive beaches and lakefront parks all along its shoreline. The Sheboygan County Historical Museum at 3110 Erie Avenue is well worth a side trip into the downtown area.

A few miles south of Sheboygan via Interstate Highway 43 and County Road V is Kohler-Andrae State Park, which preserves a full set of dune environments from newly forming dunes on the beach to stately old dune forests farther inland (Figure 4.3). A set of cordwalks (boardwalks lying on the sand) take hikers through all of these ecosystems. The Sanderling Nature Center on the lakeshore and the Creeping Juniper Nature Trail Loop provide excellent information about dune formation and dune environments. Centuries ago, the dunes blocked the Black River, which flows from the west, diverting it north around the park area toward Sheboygan. In the process, a rare dune wetland ecosystem formed here, called the Black River Marsh. It hosts remarkable biodiversity and can be accessed via a boardwalk trail into the marsh.

4.4 The old dolomite quarry at Harrington Beach State Park is now a lake.

Continuing 17 miles south of Kohler-Andrae on Highway 43 and exiting onto County Road D will bring you to Harrington Beach State Park. Here is one of the few places where Devonian dolomite is exposed. It makes up the top 10 feet of the walls of an old stone quarry in the park (Figure 4.4). Most of this rock was submerged by the lake that filled the quarry, but the top of the quarry rim is above water and can be viewed from a hiking trail around the lake. Devonian dolomite is light brown and gray, in contrast to the light gray and white Silurian dolomite.

Continue south about 30 miles on Highway 43 until you reach a number of lakeside parks north of Milwaukee, including Doctors Park on Fox Point—an expanse of green space on a bluff with access to a sandy beach along the shore. Adjacent to this park is the Schlitz Audubon Nature Center, a sprawling pre-

4.5 Lakeshore State Park in Milwaukee. WILL MARTINELLI-SPOOLMAN

serve containing wetlands, woodlands, prairie, ponds, and a lakeshore terrace sculpted by Lake Nipissing 5,000 years ago. Six miles of hiking trails and a 60-foot observation tower afford closeup and bird's-eye views of the entire area. This outdoor education center is a popular destination for hikers, bird-watchers, and students of all ages.

Another popular park just north of downtown Milwaukee is Lake Park, designed by Frederick Law Olmsted and established in 1889. It straddles the bluff, allowing for exploration of the bluff top, the beach, and broad stretches of green open space between the beach and bluff. The park also hosts wooded ravines, an Indian burial mound, a waterfall, and hiking trails.

Perhaps the most spectacular park in the city is Lakeshore State Park, located on a large peninsula just offshore from downtown Milwaukee (Figure 4.5). This park includes a large restored prairie, a lagoon with pebble beach and fishing pier, interesting sculptures, popular walkways and broad green areas to rest and picnic, wide-open views of the Milwaukee skyline, and, of course, easy access to the shore of the big lake.

Downtown Milwaukee lies in a valley created by the confluence of three rivers. The Milwaukee River flows from the north for 30 miles within one of the lowlands lying between Lake Border moraines. The Menomonee River flows in from the west through a gap in a moraine and empties into the Milwaukee River. The Kinnickinnic River flows from the southwest and joins the Milwaukee where the two rivers veer east and flow into Lake Michigan just south of downtown. The Potawatomi established a village near this site in the mid-1600s. The name Milwaukee is thought to have come from an Algonquian word meaning "good earth" or "rich, beautiful land."[13] Much of Milwaukee lies in the broad lowland stretching west for about three miles in the Menomonee River Valley. During some higher stages of Lake Michigan, the city would have been underwater on the floor of a broad inlet on that ancient lakeshore, the remnant of which is now Milwaukee Bay.

The first evidence of tropical reefs in all of North America was found just west of Milwaukee in two mounds of dolomite that contained an abundance of telltale Silurian fossils. In the late 1830s, Increase Lapham, Wisconsin's first scientist, studied these fossils and shared them with renowned paleontologist James Hall, inviting Hall to travel from New York to view the mounds. After a period of study, Hall in 1862 identified the Milwaukee dolomite mounds as cores of ancient reefs that have since been confirmed as part of the Racine Formation. They were the first ancient reefs ever identified in North America and among the first ever found in the world.[14] The dolomite in and around these reef cores was quarried in the mid-1800s for a type of stone that was used to make distinctive cream-colored bricks for many buildings in Milwaukee (the reason Milwaukee has been called the Cream City). Ironically, had it not been for the quarrying that almost destroyed the old reef cores, the reefs might never have been discovered.

Limited exposures of these reefs are accessible today via a side trip inland from the shore. The Soldiers' Home Reef is located on the grounds of the Zablocki Veterans Affairs Medical Center, which is adjacent to American Family Field, the Milwaukee Brewers' ballpark. The reef is a 35-foot-high, 450-foot-wide mound, thought to be the base of what might have been an even higher reef. It can be viewed from the ballfield's Fingers Lot off Frederick Miller Way. The ancient reef flanks the western edge of the lot on the northwest side of the ballpark.

The other reef, called the Schoonmaker Reef, after the Schoonmaker Quarry where it was discovered, is in the Milwaukee suburb of Wauwatosa. It lies along the north side of West State Street between North 62nd and North 68th Streets. It can be viewed from parking lots two blocks north of West State Street—a massive 60-foot-high, 600-foot-wide rocky bluff. The lower 20 feet of this mound is a reef core, and the upper two thirds of the hill are glacial sediments. Together, the Soldiers' Home Reef and the Schoonmaker Reef are, according to geologist Donald G. Mikulic, "the reefs that made Milwaukee famous."[15]

In Milwaukee, a trip to the Milwaukee Public Museum at 800 West Wells Street is strongly recommended. The museum has colorful and informative displays on Wisconsin's Silurian reefs (Figure 4.2), fossils from other periods, the glaciers and how they affected the state's landscapes, and many other aspects of Wisconsin history. One of the world's best collections of Silurian fossils, called the Greene Collection, is on display during limited hours in Lapham Hall on the UW–Milwaukee campus on Kenwood Boulevard. Businessman Thomas A. Greene collected a great number of the fossils in the 1850s and built his own museum, which has since deteriorated. The collection is now housed in the university's geology department.

State Highway 32 winds through downtown Milwaukee and then south along the lakeshore, providing access to more city parks—notably Sheridan, Warnimont, and Grant Parks—places to turn your back on the city for a while, enjoy the beach, and consider the vastness of Lake Michigan. From Grant Park in South Milwaukee, take Highway 32 south about seven miles to where Highway 31 splits off on a fork to the right and continues south. Follow 31 another four miles to where it meets Highway 38 (Northwestern Avenue), turn left, and drive a short distance to Quarry Lake Park in the city of Racine. This park is located on a spring-fed lake that occupies an abandoned limestone quarry with a sand beach on its south side. All other sides of the lake are bound by the vertical walls of the old quarry. In the center of the eastern wall is a massive reef core with layers of dolomite slanting off its flanks. From the beach you can see a cross section of this structure, especially the rock layers deposited on the south flank of the reef core nearest the beach.[16]

From Quarry Lake Park, follow Highway 38 (Northwestern Avenue) south and east about three miles to get back on Highway 32. From the junction, it is another 12 miles south on 32 to downtown Kenosha. The Kenosha Public Museum at 5500 First Avenue houses exhibits on geology, natural history, and fine arts. One of two wooly mammoth skeletons found in Kenosha County resides at the museum. Archaeologists have found—by studying the bones and arrowheads discovered nearby—that the animal died about 12,300 years ago and might have been killed by Paleo tradition hunters living within 200 miles of the receding glacier. Kenosha has a nicely developed waterfront, adorned by green spaces near the museum.

A good place to end the tour of Lake Michigan's shore is at Chiwaukee Prairie State Natural Area, six miles down Highway 32 from Kenosha. Its name is a combination of Chicago and Milwaukee. This preserve is located on a gentle ridge-and-swale formation that developed as the lake level dropped over the past 5,000 years. Because Lake Michigan moderates the weather, this land hosts a rich mosaic of natural communities—a sample of what the shore area was like before it was settled and then urbanized. The state natural area is mostly prairie, but it also hosts wet prairie, oak openings, and stabilized sand dunes. More than 400 species of plants have been identified there, 26 of which are rare and 10 threatened or endangered. It serves as breeding grounds for more than 75 species of grassland and wetland birds. Chiwaukee Prairie State Natural Area is also federally recognized as a National Natural Landmark.[17]

THE FOX RIVER AND LAKE WINNEBAGO

About 17,000 years ago, a great icy lake lay across much of present-day eastern Wisconsin. Shaped like a hawk in flight (Chapter 3, Figure 3.3), it sprawled across parts of 15 counties, its wings spreading from southern Columbia County to southwestern Marinette County and its tail reaching into central Fond du Lac County. The Green Bay Lobe of the glacier was retreating northeast, its meltwaters flowing in broad sheets and streams and pooling to form this vast, shallow lake, bound by highlands to the southeast, south, and northwest and by the withering wall of ice to the northeast.

Over the next four millennia as the climate oscillated between cooling and warming, the glacier readvanced and retreated at least twice. The lake, in turn, shrank away with each advance and then grew again with each glacial retreat. When the ice finally retreated for the last time, it opened an outlet that drained most of the lake, leaving a broad, muddy landscape where water flowed in meandering streams and pooled in shallow lakes. The largest of those remnants of Glacial Lake Oshkosh are now called Lakes Poygan, Winnecomb, Butte de Morts, and Winnebago, collectively called the Winnebago Pool. One of the rivers

4.6 The Fox River meanders north near where it is joined by the Portage Canal.

winding among and through those lakes and draining the nearly flat, saturated land was to become today's Fox River (Figure 4.6).

The southern half of the bed of Glacial Lake Oshkosh is now occupied by the Upper Fox River, which flows from its source to those lakes. The northern half contains the Lower Wolf River.[18] As the Green Bay Lobe and the glacial lake finally shrank away, meltwaters carved the Lower Fox River in a channel to the east of the glacial lake bed running from the north end of Lake Winnebago to the river's mouth at Green Bay. The meandering Upper Fox is 142 river miles long (covering a linear distance of about 40 miles), and the much more directly flowing Lower Fox is 39 river miles long covering about the same linear distance. Connecting the two river segments is Lake Winnebago, about 30 miles long by 10 miles wide, Wisconsin's largest inland lake, with a maximum depth of just 21 feet.[19]

The bedrock under the Fox River Valley and Winnebago Pool is dolomite deposited in Ordovician time. Late in that period, after the Appalachian Mountains were uplifted in the east by geologic forces, silt and clay washed off those then-barren highlands into the continental sea that lapped against their western flank. Currents took these sediments westward, spreading them across the continent about 450 million years ago and depositing them on the seafloor to form deep layers of soft rock, called Maquoketa shale, which after a long period of erosion, formed the lowland containing the Fox River Valley.

Before the glaciers came, there was likely a system of streams in the lowland flowing north-northeast, whittling at the shale and carrying some of it downstream, ever so gradually widening and deepening the stream beds. The glaciers worked much more quickly than the streams and removed most of the Maquoketa shale from the lowland, leaving a layer along the east and west sides of the valley above its newly plowed dolomite floor. To the east lay a gleaming white escarpment of glacially sculpted dolomite. To the west, the land sloped gently up and out of the valley. Late advances of the glacier had plastered a reddish-brown, clayey, sandy till across much of the Green Bay–Lake Winnebago–Fox River lowland. The last glacier had carried this till south from the iron-rich Lake Superior basin.[20]

Glacial Lake Oshkosh existed for 4,000 years or more, its waters collecting sand, silt, and clay and sifting these sediments onto the lake bottom. Geologists have found that thick deposits of these sediments occur in couplets, called varves, each consisting of a thin layer of coarse-grained silt and a thin layer of fine-grained clay. The former was deposited in the open lake each year during ice-free months, and the latter settled out of the lake's water during winter months when ice covered the lake's surface and currents were stilled. Geologists have found hundreds to thousands of varves, which indicate long periods of time when Glacial Lake Oshkosh sat at certain levels.[21]

The lake's shape and levels changed frequently. At its highest level—some 65 feet higher than today's Lake Winnebago—the glacial lake found an outlet at its southwest corner and drained southwest through the present-day site of Portage into the Wisconsin River until sometime after 17,000 years ago. As the Green Bay Lobe receded, it opened a series of four successively lower outlets on the lake's eastern shore, namely the channels of today's Manitowoc, Neshota–West Twin, Kewaunee, and Ahnapee Rivers, in that order. Evidence indicates that as the glacier shrank, each of these four drainage routes was short-lived, perhaps only lasting a century or two. When the glacier and glacial lake readvanced at least twice (around 15,000 and 13,500 years ago), the southwest outlet near Portage was reactivated, and it was abandoned again whenever the glacier and lake retreated.[22]

A major component of the legacy of Glacial Lake Oshkosh is the Fox River, which rises in a marshy area just north of Friesland in Columbia County. It first flows northeast, then arcs around to flow southwest for several miles to the city of Portage where it veers north again. Here is where the south-flowing Upper Wisconsin River passes within two miles of the north-flowing Fox River. Portage is so named because, for centuries, Native Americans, followed by European explorers and fur traders, portaged their canoes between the rivers, creating a vital trade route connecting Lake Michigan to the Mississippi River. Later, it became the site of a shipping canal.

The noted Wisconsin geographer Lawrence Martin speculated that someday, the Fox River might "capture" the Upper Wisconsin. Even today, heavy rains can inundate the marshy portage area and effectively join the two rivers temporarily. At the portage, the Fox is three feet lower than the Wisconsin and has a slightly steeper gradient, which gives it a fractionally faster flow. Hence, it is thought that with sustained flooding, the Wisconsin could gradually be diverted to that path of least resistance, cutting a new channel through the wetlands in and upstream of the portage area and leaving the Lower Wisconsin River downstream to become the headwaters of a newly configured river. This pirating of rivers, as some call it, is a common geological process. Rivers no less prominent than the upper Mississippi have been diverted in this way.[23] To try to prevent a diversion of the river, landowners in the 1890s built levees in this area on the banks of the Wisconsin River. The levees have since been maintained by the Wisconsin Department of Natural Resources.

The flatness of the Fox River Valley largely determines the nature of the river (Figure 4.6). The Upper Fox drops only 37 feet—less than half a foot per mile—which makes it a sluggish, meandering stream. It doesn't have enough power to erode natural blockages of glacial till, so it backs up to form wide, shallow stretches such as Buffalo Lake (12 miles long and less than a mile wide) and Puckaway Lake (seven miles long and 1.5 miles wide). Because the flat land drains slowly, it contains numerous marshes and swamps, which absorb excessive rain and act as a natural flood control for downstream areas. Puckaway Lake is shallow—no deeper than five feet—because the Fox River is gradually filling the lake with mud and other sediments. Some maps show that the lake basin was once possibly twice as large as it is today. Its west end, where the river enters, and its east end are now becoming marshland areas.[24]

Maps also show what appears to be Puckaway Lake's twin, Green Lake— another oblong lake lying just northeast of Puckaway (Figure 4.7). The two are

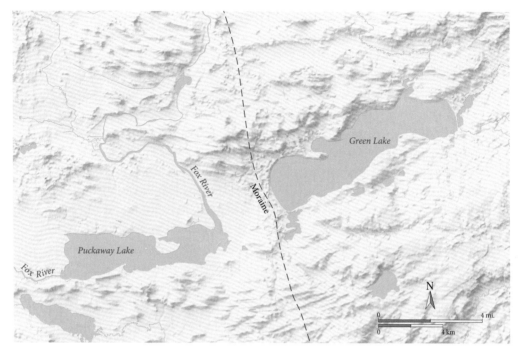

4.7 The Fox River flows through Puckaway Lake and is diverted northwest by a moraine separating that lake from Green Lake. MAPPING SPECIALISTS, LTD., FITCHBURG, WI

certainly not identical twins, for Green Lake is Wisconsin's deepest inland lake, with a maximum depth of 236 feet and an average depth of 100 feet. However, at one time, Puckaway and Green Lakes may have been much more similar or even two parts of one lake. They are separated by a 50-foot-high moraine that caused the Fox River to veer northwest instead of continuing northeast through Green Lake. After the glacier created the moraine, and when the ice started receding again, huge volumes of sediments flowed west off the melting glacier into the Puckaway Lake basin. That outwash, and the sediments carried for centuries by the Fox River, have filled this lake basin with up to 330 feet of gravel, sand, and mud. Beneath these sediments is the rock floor of a basin that, were it not for the glacier's work, would likely contain one long deep lake where Green and Puckaway Lakes now lie.[25]

Shortly downstream of where the Fox River veers northwest out of Puckaway Lake, it turns again and flows generally northeast all the way to Lake Butte des Morts, just west of Lake Winnebago. Northwest of this lake is Lake Poygan, the eastern third of which is called Lake Winneconne. Lake Poygan is so shallow that, according to some accounts, it was possible to cross it on horseback 100 years ago or more by using its sandbars as a bridge from the north shore, where the Wolf River enters the lake, to the south shore where they could access the Fox River and the other lakes.[26]

From Lake Butte des Morts, the Fox flows into Lake Winnebago near the center of its western shore at the present-day location of Oshkosh. This lake was formed by a natural dam of glacial till dropped at its northern end. Because the lake bed slopes gently eastward, the lake's west shore rises gently up into a marshy area, while much of the east shore rises precipitously up onto the Niagara Escarpment. Along the northeast shore, the escarpment has been measured as high as 313 feet above the lake level. This section's travel guide has more details on Lake Winnebago.

The Lower Fox River begins on Lake Winnebago's northwest corner, quickly arcing from northwest to northeast to complete its journey to Green Bay. Here the Fox changes from a docile, meandering stream to a more directly flowing river. Most of it flows east-northeast across bedrock that was gouged by the most recent glacier to make it more sharply sloping than it had been. The Lower Fox drops 185 feet in 37 miles, which makes its rate of fall more than 10 times that of the Upper Fox.[27] This accounts for its several sets of rapids and small waterfalls, now all underwater due to the lock-and-dam system built to make the river a shipping route. After Glacial Lake Oshkosh had drained away, the Lower Fox made its channel by cutting down through 50 to 60 feet of glacial lake deposits and creating a gorge through this sandy, clayey soil.[28] Before the river was dammed, it provided a glimpse of the bedrock buried so deeply by the glacier and glacial lake.

People have lived in the Fox River Valley for at least 9,000 years, and over the millennia, they have fished its waters, hunted its waterfowl and game, and gathered its wild rice. The Lower Fox River in particular, where archaeologists have found evidence of prehistoric and archaic villages, was occupied for centuries by ancestors of the Menominee and Ho-Chunk.[29] When European explorers first arrived in the area, one resident nation was the Meskwaki, who were given the name Renards (Fox) by French explorers Louis Joliet and Jacques Marquette, who dubbed the river Riviere aux Renards, or River of the Foxes.[30] (Today, the nation prefers Meskwaki.) Lake Winnebago once had a thriving population of lake sturgeon, and the lakes of the Winnebago Pool were important sources of wild rice. These two food sources have long been dietary staples for Native people in the area.

For many centuries, the Fox River and Lake Winnebago were part of an important Native American trade route. To connect the vast regions of the Great Lakes and the Mississippi River, travelers had only to portage fewer than a dozen rapids on the Lower Fox and to cross the marshland between the Fox and Wisconsin Rivers. Beginning in the 17th century, European explorers and fur traders followed the same route. It is thought that sometime in the 1700s, Indigenous people or fur traders carved a shallow ditch through the marshland to float their canoes across the portage, possibly using much the same route that engineers used in the 20th century to dig the canal at Portage connecting the Fox and Wisconsin Rivers.[31]

TRAVEL GUIDE
The Fox River and Lake Winnebago

The first town on the Fox, some 20 river miles downstream of the headwaters, is Pardeeville, in which Chandler Park is situated on a widened stretch of the river. This large green space is a good place to begin a tour of key points along the Upper Fox River. From Pardeeville, take State Highways 22 and 33 north and west, respectively, to Portage, where on a terrace of the Wisconsin River, the Portage Canal departs to the northeast, connecting to the Fox River (Figure 4.6). A segment of the Ice Age National Scenic Trail lies parallel to the canal, first on city streets, then through the marshy area northeast of the city where several hills are visible from the trail. Almost all of them are drumlins shaped by the glacier as it advanced from the northeast. The segment is 2.8 miles long and roughly parallels the route of the ancient portage, traveled for centuries by Native Americans, which lies a little southeast of the trail.[32]

On the east side of Portage, County Road F departs north from Highway 33 and, within three miles, passes through a marshy area. The hills on both sides of the road are drumlins, some forming islands in the broad marsh. At the end of this stretch, the road rises up onto a low ridge, which is a moraine left by the Green Bay Lobe of the glacier, and about a mile farther, Highway F curves around a large drumlin just before its junction with Fox River Road. Take this road north a short distance and turn left on Lock Road to travel west to Governor Bend Park on the Fox River—the site of an abandoned lock-and-dam, now a pleasant green space on the river. Here, the river valley has narrowed to about a quarter of a mile wide. For many centuries, Glacial Lake Oshkosh drained through this valley into the Wisconsin River. This channel would have been filled brim to brim with icy meltwater during much of that time.[33]

Back on Fox River Road, continue north to rejoin Highway F. Three miles beyond this junction is John Muir Memorial Park on Ennis Lake. Here on a rocky moraine is where Muir's father built a farm from which young John began his storied career as a naturalist and writer. A segment of the Ice Age Trail loops around Ennis Lake, providing beautiful views of this pristine body of water and retracing the steps Muir probably took as a boy beginning his lifelong exploration of wild places.

One way to experience the presettlement Fox River environment is to visit Page Creek Marsh State Natural Area. From Muir Memorial Park, take Highway F a short distance north to its junction with 10th Road, which runs northwest for about seven miles to County Road D. Cross D and continue north on County Road K a short distance to a secluded parking area on the right side of the road shortly before K turns east along the shore of Buffalo Lake, which is a widening of the Fox River. This wetland preserve on Buffalo Lake hosts a sedge meadow, oak savanna, prairie, bogs, and small lakes, and an outstanding diversity of plants, meadowland birds, and waterfowl. It is also an important staging area for sandhill cranes at fall migration time.[34] This hike is relatively challenging because of the thick vegetation that can obscure the trail. However, the land is owned by the Nature Conservancy, which has preserved the ecosystem in a close-to-presettlement condition.

County Road K continues along the south shore of Buffalo Lake to the town of Montello, famous for spectacular outcroppings of red granite, which were quarried for nearly a century beginning in the 1880s (Figure 4.8). In Montello, the river swings from its eastward flow to the southeast, as if heading back toward its headwaters. It crosses a broad wetland area and, having flowed more than 80 river miles from its headwaters, enters Puckaway Lake, the east end of which is just eight miles north of the river's source. This is a typical flatland river, meandering to meet itself at least once.

The Fox River next winds its way to the town of Princeton and then, eight miles farther downstream, skirts the huge White River Marsh State Wildlife Area. To get there, drive about 10 miles northeast from Montello on Highway 23 to Princeton, and there, use West Main Street to jog onto River Road (County

4.8 Montello's famous red granite is exposed here in an old quarry on the Fox River.

Road D) and take this road north about three miles to White River Road to access the state wildlife area. This 12,000-acre area contains open marsh, wet meadows, hardwood and tamarack swamps, prairie, and oak savanna—another sampling of presettlement Fox River environs. The southwest corner of the marsh, closed from mid-June to mid-October, is used for training and releasing whooping cranes in an effort to help restore their populations. Elsewhere in the White River Marsh State Wildlife Area, visitors can hike, canoe, and enjoy excellent wildlife viewing opportunities accessed from parking areas along County Roads DD and D and White River Road.[35]

From here, the Fox River ambles northeast to the town of Berlin and then flows more directly east-northeast through a mix of farmland and wetlands to Lake Butte des Morts where it is joined by the Wolf River, which flows from the north. The Fox River then flows through the city of Oshkosh and on into Lake Winnebago. From White River Marsh, the river is harder to follow by road. To get to Oshkosh, take White River Road north out of the marsh to County Road D, continue north to County Road F, and turn east to get to Berlin. From there, take State Highways 91 and 44 to Oshkosh. The Oshkosh Public Museum at 1331 Algoma Boulevard features "People of the Waters," an interactive display of hundreds of artifacts representing 13,000 years of human history in the area from the earliest people through the fur trade years.

To drive around Lake Winnebago, take US Highway 45 south from Oshkosh along the gently sloping western shore of Lake Winnebago all the way to the city of Fond du Lac, whose French name translates to "bottom of the lake" or "foot of the lake." At the city's picturesque Lakeside Park, pick up Lake Winnebago Drive, which angles northeast along the shore and connects with US Highway 151 as it continues north along the east side of the lake. This road lies away from the much steeper east shore and runs along the top of the Niagara Escarpment. It affords panoramic views of the lake, especially at waysides along the highway. Allow your imagination to wander back about 17,000 years to when Glacial Lake Oshkosh sprawled into the distance, far out of sight. That ancient lake would have been littered with icebergs calving off the wall of ice that was retreating to the north.

About halfway up the east shore, when Highway 151 turns east, stay north on State Highway 55 as it continues along the top of the Niagara Escarpment. Follow 55 through its junction with State Highway 114 as it runs around the northeast end of Lake Winnebago through the village of Sherwood. Bear left on 114 and then turn left on Pigeon Road, which goes south to State Park Road and High Cliff State Park. This park provides the best views available of the Niagara Escarpment along Lake Winnebago. The dolomite cliff here, locally called "the Ledge," rises to more than 210 feet above the lake and stretches for five miles from Sherwood south to just north of Stockbridge. Hikers can get spectacular, broad views of Lake Winnebago as well as closeup views of the layers of ancient dolomite that make up the escarpment (Figure 4.9). The park also has a well preserved set of effigy mounds, probably built by people of the Woodland tradition between 500 and 1,000 years ago.

Beyond the park area, Highway 114 continues for seven miles to the city of Menasha. Anyone interested in geology must visit the Weis Earth Science Museum, designated "the official mineralogical museum of Wisconsin," on the UW–Oshkosh Fox Cities Campus at 1478 Midway Road. Along with an impressive array of minerals and fossils, the museum boasts interactive exhibits

4.9 A section of the Niagara Escarpment on display in High Cliff State Park

that allow visitors to experience an earthquake, walk through a mine tunnel, touch dinosaur bones, and make it rain. Also on display is an impressive wood carving of a contour map of northern Lake Winnebago. The museum's mission is to "preserve and promote the unique geologic heritage of Wisconsin and to underscore the importance of earth science to society . . . through education, research, and through the collection, conservation, and display of minerals, rocks, fossils" and other historical objects.[36]

The Fox River flows briefly west from Menasha and Neenah, then curves northeast to flow through the cities of Appleton, Little Chute, Kimberly, Combined Locks, and Kaukauna. There follows a 15-mile, relatively less urbanized stretch of the river. You can see much of the Lower Fox River by taking State Highway 47 north to Appleton from its intersection with 114 in Menasha. In Appleton, turn right onto State Highway 96 and travel to Wrightstown. Then follow County Road D the rest of the way to De Pere, one of Wisconsin's oldest cities, just south of the city of Green Bay.

In Green Bay, where East Mason Street crosses the river, is the northern trailhead for the Fox River State Recreational Trail, which runs for five miles along the river before curving onto an old railroad grade running south through several Brown County communities. Also in Green Bay, on the east bank of the Fox River at the intersection of Webster and Greene Avenues, is Heritage Hill State Park. This is a living history museum comprising several historic buildings and thousands of artifacts and using costumed interpretors to portray people from all phases of northeastern Wisconsin's early and modern history.

Horicon Marsh

A dawn wind stirs on the great marsh. With almost imperceptible slowness, it rolls a bank of fog across the wide morass. Like the white ghost of a glacier, the mists advance riding over phalanxes of tamarack, sliding over bogmeadows heavy with dew. A single silence hangs from horizon to horizon.

—Aldo Leopold

The great marsh that Aldo Leopold venerates in "Marshland Elegy," an essay that appears in his classic *A Sand County Almanac*, is the Horicon Marsh.[37] It is indeed a wetland to be venerated, one of the largest freshwater cattail marshes in the

4.10 Horicon Marsh on a summer morning. The forested low mounds in the distance are drumlins, which have become islands in the marsh.

world, and host to a spectacular biodiversity of wetland plants and animals. It occupies 31,000 acres, most of it in Dodge County with its northern tip lying in Fond du Lac County. Roughly oval-shaped, it is 13 miles long and up to 4 miles wide (Figure 4.1). The marsh is fed by the headwaters of the Rock River—three branches that drain 300 square miles of land, funneling surface water into the marsh. The great marsh owes its existence to the glacier that Leopold imagined seeing one morning in that ghostly bank of fog.

While most of the glacial deposits that lie under the marsh are between 12,000 and 16,000 years old, the story of how the marsh was formed is much older. It rests in the same lowland that contains Green Bay and the Fox River Valley, so the bedrock beneath it is composed of Maquoketa shale overlying layers of Ordovician dolomite.[38] During the Ice Age, the lowland was carved deeper with each passage of a glacier over the area. Most of the work was accomplished during the Wisconsin glaciation, but geologists have found evidence that earlier glaciers helped to shape the land. A buried channel beneath today's East Branch of the Rock River, which flows from the east into the marsh, likely carried melt-water from one of those earlier ice sheets. And deep layers of silt and clay in the Horicon Marsh make it clear that a large lake occupied the marsh basin before the Wisconsin glaciation.[39]

The most recent glacier scooped the basin deeper than it had ever been and shaped the landforms that today lie within the marsh. The marsh's islands are drumlins carved by the glacier as it advanced. Some of these, including Stony Island, Goose Island, and Misling Island, are made largely of bedrock that was scraped and molded into the typical teardrop shape by the moving masses of ice. Others, including Phil's Island, Eagles' Nest, and the Palmatory Drive drumlins, are made of sand and gravel that were frozen hard under the glacier and thus resisted erosion enough to remain first as hills, now as islands. The large peninsula of higher land protruding into the northeast corner of the marsh is not a drumlin, but it was also molded by the glacier. It is a mass of Ordovician dolomite that has resisted erosion.[40]

At its peak, the Green Bay Lobe of the glacier completely covered the Horicon Marsh area. As it receded, it halted at various times and sat for many years, creating recessional moraines. One of them—the Green Lake Moraine—was formed along a line running east to west through the present-day city of Horicon, a little south of West Lake Street. The ice then receded to just north of the marsh, and there it rested again, forming a series of low ridges called the Waupun Moraine between 12,900 and 12,000 years ago. During that time, the area was a frozen tundra that likely hosted wooly mammoths who ranged the hills and munched on the vegetation. They are memorialized today on the grounds of the Horicon Marsh Education and Visitor Center by an iron sculpture that stands like a sentinel on a rise overlooking the marsh (Figure 4.11).

The melting mass of ice produced huge volumes of meltwater that carried the silt, sand, and gravel picked up by the glacier. Some meltwater streams cascaded off the top of the glacier onto newly exposed land, while others flowed under the ice. Where the latter streams exited the ice mass, they spread glacial debris in fan-shaped outwash plains, one of which is prominent today in the northeast corner of the marsh.[41]

The Waupun and Green Lake Moraines bracket the marsh basin, forming its north and south ends, respectively. Here, in the lowland defined by the moraines and by the higher land on either side of the marsh, meltwater collected over several centuries, forming Glacial Lake Horicon. Evidence suggests that

4.11 This true-to-scale iron mammoth surveys the marsh on a summer morning, sharing the view with the author's wife, Gail.

abundant flows of water moved into and out of the marsh basin. To the north is a buried ancient channel up to 600 feet wide that once carried water off the receding glacier and into the glacial lake. The South and East Branches of the Rock River are ancient streams that flowed along the margin of the glacier while the Waupun Moraine was forming. The East Branch brought water down from the Niagara Escarpment when it was still largely covered by ice.[42]

These ancient streams carried huge volumes of silt and clay into the basin, beginning about 10,400 years ago, and modern streams continue to do so. Siltation has been a major force in the formation of the Horicon Marsh, accumulating to depths of up to 60 feet in parts of the east side of the marsh. Ancient Lake Horicon originated in a smaller basin on the east side of today's basin, and then spread west, partly because it was filling with silt. As the lake grew, waters filtering south through the Green Lake Moraine eroded a notch and eventually breached the moraine. As the newly formed river drained the ancient lake around 2,470 years ago, it carved the channel of the modern Rock River and converted the lake to a vast marsh. Today, ancient beach lines and terraces in the marsh show this progression from lake to marsh.[43]

That progression was a long and apparently cyclical process. Geologists have drilled into the marsh, bringing up core samples that tell of ancient wetland vegetation growing at least 10,300 years ago. It lies under layers of clay and silt from meltwater streams interspersed with layers of more recent cycles of vegetation. Throughout many centuries, deposits of peat also accumulated in the marsh after the drainage of the lake. Over thousands of years, clay, silt, and peat created fertile soil for the proliferation of aquatic plants and the animals that feed on them, adding to the rich glacial legacy of the marsh area.[44]

Eventually, forest species took hold on the islands and other higher, drier areas in and around the marsh. The progression was typical of postglacial forest development, beginning with spruce and gradually shifting to lowland hardwoods, including several species of oak, ash, and cherry trees. Ground cover included various fern species, as well as wildflowers such as white trillium, Dutchman's breeches, mayapples, and wild geraniums. Today, the limited forest stands in the area provide cover and food for migrating warblers and resident woodpeckers, among other woodland bird species. Some of the islands have also hosted thriving heron rookeries.

In the wetter parts of the marsh, aquatic plant and animal diversity also grew steadily whenever the marsh was undisturbed for many years, and today, cattails, bulrushes, sedges, and smartweeds grow densely and support redwing blackbirds, ducks, and many other migratory species. American white pelicans and great blue herons use the islands and interior of the marsh for nesting and feeding. The pelicans work in groups, herding fish into shallow water to maximize their catches. Other favorite species for bird-watchers in the marsh include abundant Canada geese, several duck species, sandhill cranes, tanagers, orioles, red-tailed hawks, eagles, and many more. All told, the entire marsh area provides critical habitat for more than 300 species of birds, as well as red fox, muskrats, bats, turtles, frogs, fish, and dragonflies.[45]

The marsh's diversity also includes, and is threatened by, invasive species such as carp—which uproot plants, stir sediments, and cloud the waters—and purple loosestrife, which outcompete native plants along watery margins in the marsh. Wild parsnip is a threat in some upland prairie areas around the marsh, and European buckthorn and garlic mustard threaten to crowd out diverse native ground cover plants in certain forest stands.[46] All of these non-native species can disrupt food webs and thus do great harm to the ecosystems they invade.

Horicon Marsh's natural history includes its use by humans, beginning around 12,000 years ago. Nearly every major prehistoric culture known in the Upper Midwest is thought to have used the marsh, including the Paleo and Archaic cultures, the Mound Builders, and other Woodland cultures, and more recently the Hopewell culture. The word *Horicon* is derived from an Algonquian term meaning "land of clean, pure water."[47] Historically, the east side of the marsh area was occupied primarily by the Potawatomi and the west side by the Ho-Chunk.

Archaeologists have found evidence of a network of trails around the marsh, used for centuries by Indigenous people. A number of well-traveled trails met at the south end of the marsh in a village and ceremonial grounds were located where the city of Horicon is today. Four of these trails once lay where modern highways now run: County Road A from Horicon to Fox Lake, State Highway 26 on the marsh's west side, State Highway 33 on its south end, and County Road Z along the east side.[48]

Effigy mound builders were present in the Horicon Marsh area between 700 and 1200 CE. When Wisconsin's first scientist, Increase Lapham, surveyed the area in the 1850s, he found more than 500 such mounds around the marsh. Most of them have been destroyed for farming and other reasons, but some, thanks to Lapham's early efforts, have been preserved. One of the best surviving clusters of mounds in the state can be viewed at Nitschke Mounds County Park near the southwest corner of the marsh, at N5934 County Road E. This archeological site contains 39 effigy mounds thought to have been built between 800 and 1200 CE by Late Woodland people. (See travel guide.)

With the arrival of Europeans in the 1800s, the story of Horicon Marsh departs from natural history and enters a tortured period of disturbances that greatly degraded the marsh ecosystem. It began in 1846 when the town of Horicon was founded by white settlers who built a dam on the Rock River at Horicon to facilitate lumbering and farming. This created a modern Lake Horicon, which reached depths of up to nine feet. In 1869, developers who wanted to restore the marsh destroyed the dam and drained the lake. Waterfowl flocking back to the restored marsh became the new target of exploitation. Hunters, too, flocked to the marsh and, between the 1870s and early 1900s when hunting was unregulated, nearly wiped out the marsh's duck and other waterfowl populations.

Next in the story of the marsh's degradation was a short period of attempts to farm the marsh, first by draining it between 1910 and 1914. For a few decades, farmers tried and failed to cultivate the drained land at a great cost to the marsh ecosystem. In draining large tracts of the marsh, they exposed peat beds, which eventually dried and caught fire. The burning of the marsh left it useless for farmers and nearly uninhabitable for wildlife. Drainage ditches scarred the land, and some are still visible today. The rich deposits of peat, up to 18 feet deep in places, had been converted to muck by 40 years of draining, failed farming attempts, and burning.

Most of this degradation was eventually reversed, beginning in 1921 with efforts by concerned citizens to begin restoration. They got a boost in 1927 with the passage of the Horicon Marsh Wildlife Refuge Bill, passed by the state legislature. It led to the state's acquisition of land to be restored in and around the marsh and to the building of a dam in Horicon that restored water to the devastated marsh. In the 1940s, the US Fish and Wildlife Service purchased the northern part of the marsh to establish a federal wildlife refuge.[49] The abandonment of efforts to develop the marsh area resulted in the restoration of one of the country's largest freshwater marshes—a testament to letting nature take its course.

Today, Horicon Marsh is a destination for thousands of visitors every year, including bird-watchers who come to see a stunning array of species, especially during migrations from mid-April to mid-May and from mid-September through October. A limited amount of fishing, hunting, and trapping is also allowed in the marsh. Both the federal and state wildlife refuges have trails that lead hikers through the woodlands and grasslands to views of the restored wetlands and lakes of the marsh (Figure 4.12).

Because of its high value as one of Wisconsin's premier wetland and water destinations, Horicon Marsh has received protection under state and federal laws. The southern third of the marsh (11,000 acres) is operated by the Wisconsin Department of Natural Resources as a state wildlife area. The northern portion (21,400 acres) is protected as the Horicon National Wildlife Refuge. The marsh has also been designated as a unit of the Ice Age National Scientific Reserve—one of a set of nine sites enjoying the same protections as national parks. This designation recognizes the marsh as a prime example of an extinct glacial lake. Finally, the marsh has been recognized internationally by the Ramsar Convention of the United Nations as a Wetland of International Importance.

In 1948, Aldo Leopold expressed some pessimism about the restoration of Horicon Marsh. Observing the brutal and widespread disturbances to the

4.12 Hundreds of species of birds and other animals thrive in Horicon Marsh. PHOTO BY JACK BARTHOLMAI

marsh, he feared that someday "the last crane will trumpet his farewell and spiral skyward from the great marsh."[50] No doubt, Leopold would be pleased to see that his pessimism was unwarranted in the long run. The marsh has survived those disturbances to become a thriving marshland ecosystem.

TRAVEL GUIDE
Horicon Marsh

Horicon Marsh can be explored by canoe or kayak, on foot, by bike, or in a car. A good place to start is at the Palmatory Street Overlook on the north side of the city of Horicon. Palmatory Street runs north from State Highway 33, one block west of the junction of Highways 33 and 28. The overlook is on a pair of high drumlins that extend into the marsh. The parking area rests on top of the southern drumlin, where vantage points provide panoramic views to the west and east. A gravel hiking trail runs down the front (north side) of the southern drumlin and up over the northern drumlin—a total length of half a mile. A mowed grass trail runs along the east flank of the drumlins. At its northern end, the central gravel trail connects with a trail on a dike that crosses to the southeast side of the marsh.

That latter trail is part of a set of loop trails that depart from the Horicon Marsh Education and Visitor Center, an impressive facility run by the Friends of the Horicon Marsh, just a short drive north of the Palmatory Overlook on State Highway 28. The trails take hikers out into the marsh on earthen dikes and boardwalks that provide excellent opportunities to view the lush wetland forests, prairie remnants, wild flowers, waterfowl, and ponds teaming with amphibians that bring the marsh to life from spring through the fall. Other informative visitor facilities are the National Wildlife Refuge Visitor Center on the east-central side of the marsh and the Marsh Haven Nature Center on the north end of the marsh.

For displays on Native American history in the marsh area, visit the Horicon Historical Society's Satterlee Clark House Museum at 322 Winter Street in Horicon and the White Limestone School Museum on North Main Street in Mayville. On the southwest corner of the marsh is Nitschke Mounds County Park, described earlier in this chapter. A trail with interpretive signs runs among several mounds in the park, which is located at N5934 County Road E.

It is possible to reach all of these places by taking a drive around the marsh on the Horicon Marsh Parkway, starting on Highway 28 out of Horicon heading northeast toward Mayville; turning north of County Roads TW, Y, and Z along the east side of the marsh; and then turning east on State Highway 49 across the north end of the marsh. County Road I runs south from 49, just east of Waupun, to merge with State Highway 26 running along the marsh's west side. County Road E runs east from 26 across the marsh's south end into Horicon to complete the loop.

Dodge County Ledge Park, a few miles east of Horicon, provides a side trip with a different perspective on Horicon Marsh. The park is located on top of the Niagara Escarpment, which forms the eastern border of the broad lowland containing the marsh. From high up on the park's cliffs, visitors get sweeping views of the entire marsh. Hiking trails provide a closeup look at the craggy mass of dolomite that forms the escarpment. If you reach the Horicon Marsh area on a sunny morning, you may wish to start your day with this side trip.

The sun lighting the marsh from the east truly enhances the view. To reach the park, take Rasch Hill Road east out of Horicon about three miles and turn north on Park Road.

Yet another side trip for hikers and bikers is the Wild Goose State Trail. Lying on an old railroad grade from Clyman Junction, southeast of Horicon, to the city of Fond du Lac to the northeast, the trail runs the entire length of Horicon Marsh on its west side. It veers closest to the marsh at its north end and can be picked up there, where it crosses State Highway 49.

The Yahara Lakes

Running through the center of Dane County is the Yahara River. The valley of the Yahara is central not only to the county itself but to the story of Dane County's geology and natural history. In the most recent chapter of that story—the coming and going of the glaciers—five large basins were formed in the Yahara River Valley. Within those basins now lie the county's five major lakes: Mendota, Monona, Wingra, Waubesa, and Kegonsa—the Yahara Lakes (also known as the Madison Lakes, Figure 4.13).

The story of these lakes begins, as do most geologic stories, with the formation of the bedrock underlying them. One major component of the bedrock under the Madison area—its basement rock—is layers of Cambrian sandstone interspersed with thin layers of shale. Overlying these layers is Ordovician dolomite, which now forms the surface of the bedrock underlying most of Dane County. All of the county's younger Silurian and Devonian dolomites have been eroded away.

Rivers did much of that erosive work. Sometime before the Ice Age arrived, there probably existed an ancient version of today's drainage system, which includes the Yahara River flowing to the Rock River. As the glaciers were expanding in the north, the global ocean was dropping, its water being taken up the growing ice masses. By the time the glaciers reached Wisconsin, the global sea level had dropped 350 to 400 feet below present levels. Coastal shorelines thus receded and dropped to that level, which affected many streams and rivers. A stream tends to seek its lowest possible level, which is defined by the water level at its mouth where it stops flowing and thus stops downcutting into its bed. As sea levels and coastlines were dropping, the rivers that fed the oceans sought to

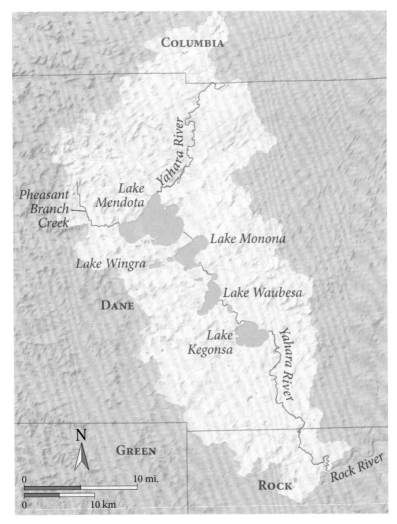

4.13 The Yahara River Watershed and Yahara Lakes. MAPPING SPECIALISTS, LTD., FITCHBURG, WI

reach those lower levels where they entered the oceans, and thus they cut their valleys that much deeper. Over time, the Mississippi River carved its valley 350 to 400 feet deeper than it is today. Likewise, its major tributaries, including the Wisconsin and Rock Rivers, cut their valleys deeper, and so did their tributaries, including the Yahara River.[51]

Just before the glaciers arrived, southern Wisconsin likely looked much like Wisconsin's Driftless Area appears today, with high, rocky hills and ridges and deep river valleys. The Yahara was probably a good-sized river sweeping through the present-day Madison area. Madison's isthmus was a long bedrock ridge, around which the river snaked as it made a grand S-turn, flowing through the future basins of Lakes Mendota and Monona. From sites like Picnic Point on Lake Mendota and Turville Point on Lake Monona, where today we gaze out at placid waters, a preglacial hiker would have looked down 250 to 300 feet at a wide river winding through the ancient valley.[52]

The Wisconsin glaciation did most of the landscaping we see now in the Yahara Valley. At its peak, sometime between 20,000 and 14,000 years ago, the Green Bay Lobe extended over the area with an estimated thickness of 1,600 feet, tapering to a few dozen feet in the western third of Dane County, where it built the Johnstown Moraine. As the climate warmed, the glacier retreated haltingly, building mostly smaller recessional moraines east of the Johnstown. The most prominent of these is the Milton Moraine, which runs through west Madison, paralleling the Johnstown and defining the western side of the Yahara River Valley.[53]

The glacier widened the river valley, picking up crushed sandstone, boulders, gravel, sand, and clay, and carrying this glacial debris to the outer margins of the ice mass. It shaped drumlins in the valley, some of which are still prominent today, not least of which is the site of the Capitol building in central Madison. However, the retreat of the glacier was perhaps equally significant in shaping the Yahara River Valley and Lakes, mostly due to the formation of glacial lakes.

Imagine a wall of ice standing on the west side of the Yahara River Valley, along a northwest-southeast line just west of Lakes Kegonsa, Wabesa, Wingra, and Mendota. It had lain there for centuries, building the Milton Moraine. During that time, meltwaters had formed Glacial Lake Middleton in a low area between the Johnstown and Milton Moraines in what is now northern Middleton, just northwest of Madison. That lake drained to the northwest through Black Earth Creek. About 15,000 years ago, the glacier began to waste back again to the northeast. Its meltwaters pooled against the ice wall just east of Lake Middleton to form the early phase of Lake Mendota, whose waters first drained west through what is now Pheasant Branch into Lake Middleton. At the time, the level of the two lakes was about 90 feet higher than today's Lake Mendota.[54]

As the glacier continued to retreat, more glacial lakes pooled between the ice wall and the Milton Moraine, forming early, shallow versions of today's lakes. The retreating ice revealed an outlet to the southeast—perhaps the bed of the ancient Yahara River—that was lower than the northwest outlet. The lakes began to drain to the southeast, at which time the ancient Pheasant Branch reversed direction and Lake Middleton began draining east into Lake Mendota.

The ice melted back in fits and starts, its withering margin remaining roughly parallel to the Milton Moraine. Along the way, it calved huge chunks of ice, which fell to the newly ice-free land and were then gradually buried by tons of glacial debris flowing away from the glacier. A number of these icebergs fell along the axis of the Yahara River Valley, and the kettles that remained when this ice melted out, probably centuries later, make up parts of the large basins that now contain the five lakes.[55] By around 13,500 years ago, the glacier had melted out of the Yahara River Valley completely.[56]

Even as the ice wall stood east of the valley, enormous amounts of sand and gravel washed off the retreating glacier to partially fill the valley's kettles and other low areas. Geologists estimate the depth of this material over the bedrock in the valley today to be more than 300 feet in many places.[57] Some of this material was shoved along and molded by rushing meltwaters, and it accumulated to form dams at certain points across the ancient Yahara River. In some cases, water flowing behind these natural dams would whirl against the riverbed, scooping outwash material from the basins and building the dams higher.

The combination of this natural damming and the voluminous meltwater flows caused the valley's small nascent lakes to merge, forming Glacial Lake Yahara. It was an estimated 12 feet higher than today's Lake Monona and was dotted with islands—the drumlins and other hills in the valley formed by the glacier.[58] Were we to turn the clock back to this time, people visiting the Capitol building today would suddenly find themselves on an island in a vast, icy lake.

Over several thousand years, Glacial Lake Yahara slowly drained to the southeast as the Yahara River cut through glacial deposits to deepen its channel. Islands and ridges emerged from the water to form a soggy landscape that defined the five lakes we know today. Geologists do not know exactly how long it took. They think that Lake Middleton persisted, slowly draining into the Yahara River Valley, long after the glacial retreat. Historical accounts tell us that by 1833, Lake Middleton was a big marsh.[59]

Similar changes have occurred elsewhere in the Yahara River Valley during the centuries after the glacier finally retreated. The shallow Lake Kegonsa is thought to have been much bigger immediately after the glacier left, probably occupying most of the marshy area now north and northwest of the lake.

Likewise, the southeast end of Lake Waubesa became a large wetland (Figure 4.14), and Lake Wingra was probably twice the size it is today, much of it being converted to wetland during those postglacial centuries.[60] But for a low recessional moraine that had been formed by the last glacier between Lakes Wingra and Monona, the two might well have been one lake with a long, narrow bay extending through a deep bedrock valley that opened into the Yahara River Valley. The glacier had truncated the bay by dropping a ridge of debris across it, leaving the new Lake Wingra lying to the west. That moraine was largely leveled beginning in the 1870s—quarried of its sand and gravel to help build the city of Madison—but it still separates Monona Bay from Lake Wingra.

The postglacial Yahara River looked much as it does now, with its course almost completely contained within what would become Dane County. Its short valley had been carved wider by the glacier and its meltwaters, which easily eroded the dolomite and sandstone bedrock and opened a two- to five-mile-wide trench. The river rises from springs flowing within a cluster of drumlins that straddle the Dane County–Columbia County line northeast of Lake Mendota. It then meanders south and southwest through broad wetlands and winds among several more drumlins before entering Lake Mendota. In the nine-mile traverse through Lakes Mendota, Monona, Waubesa, and Kegonsa, the river drops just six feet. It then flows to the Rock River a few miles south of the Dane County–Rock County line.[61] From there, the Yahara River's waters mingle with those of the Rock, which flows south out of Wisconsin, curls across northwestern Illinois, and empties into the Mississippi River.

Preglacial plant communities in central Dane County likely resembled those in the less developed parts of today's Driftless Area—old forests standing in the deep valleys with savannah or prairie occupying most of the upper slopes and ridgetops. Immediately before and soon after ice covered the region, however, it was a treeless tundra supporting only extremely cold-weather species such as lichens. Within decades of the glacier's retreat, the soggy, newly molded landscape would host a few grasses and shrubs. In stream valleys and other lower areas, black spruce were the first trees to take hold. Shallow lakes among the drumlins soon had peat bogs spreading from around their edges. Spruce forests saw other types of conifers move in. As the landscape dried somewhat with the warming climate, a mix of pine, ash, birch, elm, and oak became established in increasingly diverse forests. On the drumlins and other high areas, oak savannahs and prairie became part of the mix.[62]

4.14 The Waubesa Wetlands host a rich diversity of plant and animal communities. CALVIN B. DEWITT AND NADIA OLKER

Not long after the climate began to warm, large mammals migrated from the south into Wisconsin. Grazing mammoths and musk oxen and browse-eating mastodons wandered on the grasslands and among the spruce stands. Giant beavers more than six feet long inhabited the ponds and shallow lakes. The bones of all these animals have been found in glacial lake sediments in eastern Dane County. All of them became extinct by around 4,000 years ago.[63] Many smaller mammal species, as well as waterfowl, moved in after glacial times.

Following the large mammals were hunters of the Paleo tradition—the first humans to move into the region around 12,000 years ago. After the megafauna had gone extinct, seminomadic people of the Woodland tradition lived in southern Wisconsin, hunting other mammal species, fishing the waters, and gathering herbs, nuts, and berries in the forests. Eventually, they established villages in several locations on the lake shores and linked them with a network of trails. The Capitol building sits at the junction of several well-traveled trails.

Beginning around 800 CE, Late Woodland cultures took to growing corn, squash, and beans. Notable locations of their villages are present-day Tenney Park on the east shore of Lake Mendota and the mouth of Pheasant Branch on the northwest shore of that lake. Some Woodland groups built effigy mounds in the shapes of hawks, pumas, deer, bear, and other animals. The shores of the Yahara Lakes once held one of the highest concentrations of such mounds in the world, estimated to number more than 1,500.[64] Most of them have since been destroyed for farming, road building, and urban development.

In 2021, Tamara Thomsen, a Wisconsin Historical Society archaeologist, discovered more evidence of these effigy mound builders in the form of a 1,200-year-old wooden dugout canoe lying on the bottom of Lake Mendota. Analysis of this artifact revealed that it likely was built around 800 CE by ancestors of the modern Ho-Chunk nation. Within the ancient craft researchers found net sinkers—stones fashioned to hold down fishing nets—which might reveal more about the fishing methods and lifestyles of these Woodland mound builders.[65]

Not surprisingly, the names of the lakes come from Native American languages. But rather than being used by the area's Indigenous people themselves, the names were assigned by surveyors and prominent early settlers in Madison (among them Lyman Draper, who founded the Wisconsin Historical Society). A hodge-podge of languages is evident in the lake names, which were officially adopted by the state legislature in 1855. *Mendota*, the largest lake, is a Siouan (Dakota) word meaning "mouth of the river" or "confluence of rivers." *Monona* is believed to have come from an Ojibwe word meaning "beautiful." *Waubesa*, also Ojibwe, means "swan."[66] *Kegonza* comes from the Ojibwe, meaning "lake of small fishes," or possibly from a Ho-Chunk word meaning "lake of many fishes."[67] *Wingra*, meaning "duck" in Ho-Chunk, was called "Dead Lake" by settlers, possibly because it was separate from the other four lakes, connected by only a small stream.[68] In recent decades, the Ho-Chunk have revived the use of the place name *Dejope* (Four Lakes) to refer to Madison and its surrounding lakes.

Today, the lakeshores are dominated by urban development, the river valley by agriculture. The Cambrian sandstone deep under the city of Madison is a massive aquifer that supplies most of the city's drinking water. Between the Cambrian and Ordovician bedrock layers was a transitional stone, a dolomitic sandstone once known as Madison sandstone. It was quarried extensively on the city's west side and used to erect some of Madison's oldest buildings, including the first structures on the University of Wisconsin campus—Bascom Hall and North and South Halls, all built in the 1850s.[69]

Much of Madison is built on filled wetlands that emerged as Glacial Lake Yahara drained. The Dane County Regional Airport, parts of the UW–Madison Arboretum, and most lakeside parks lie on deep sediments deposited by the ancient lake. Other areas of the city sit on glacially formed highlands. As noted earlier, the hill under Capitol Square is a drumlin, as is Mansion Hill just north of the Capitol. A little to the west of Mansion Hill are UW–Madison's Bascom Hill and Observatory Hill, both well known to the many thousands of students who have trudged up and down those drumlins since the 1850s. All of Madison's drumlins were, at some point, islands in Glacial Lake Yahara.[70]

As for Lake Mendota, its modern story is at least as famous as its ancient history, certainly among scientists anyway, for it is one of the world's most studied lakes. It was here that the field of limnology in North America grew, beginning in the 1930s, after taking root in the 1920s at the university and at the state's Trout Lake Limnological Laboratory (see "Lake Country" in Chapter 2). Limnology is the study of inland waters—their chemistry, physics, geology, hydrology, and ecology. In the early 20th century, Wisconsin's own Professors Edward A. Birge and Chancey Juday, working at Trout Lake, founded the North American branch of the field. Later, Professor Arthur D. Hasler ran the program and expanded it in a laboratory on the south shore of Lake Mendota. Countless studies on the lake have since generated voluminous data and models for research that have served limnologists throughout the world.[71]

TRAVEL GUIDE
The Yahara Lakes

There are any number of ways to tour the Yahara Lakes. Here is one suggested plan.

Lake Mendota

To begin a circle tour of the lake, take Johnson Street in downtown Madison and head northeast on Madison's Isthmus to find two parks on the lake's southeast shore. The first is James Madison Park, a popular destination for UW–Madison students and, like the city, named for the fourth president of the United States who died in 1836 as the city was being planned by its founders. Farther up the shore is Tenney Park, a shaded green expanse centered on what was probably a meander or oxbow on the ancient Yahara River, which had flowed out of Lake Mendota at this point on the shore. The river has since been channelized to cut straight across the Isthmus. Views from both of these parks give visitors a sense of the size of this biggest of the Yahara Lakes (Figure 4.15).

North of Tenney Park on the western shore of the lake is the village of Maple Bluff, featuring spectacular Cambrian sandstone bluffs, mostly on private land but viewable from the lake for boaters exploring the shoreline. On the northeast shore of the lake is Warner Park, where North Sherman Avenue meets Northport Drive (State Highway 113). This is another broad green expanse centered on a ring-shaped waterway, perhaps a remnant of a tributary to the preglacial Yahara River.

On the north side of Warner Park, Troy Drive runs west from 113 toward Farwell's Point, where the sprawling grounds of the Mendota Mental Health Institute are accessible from the west end of Troy Drive. This is likely the site of a village from the Late Woodland period. It contains well-preserved examples of effigy mounds, including a bird mound with a wingspan of 634 feet. Also accessible from the institute's grounds is Governors Island, a 60-acre isle connected to the mainland by a land bridge made of fill. Hikers can walk its perimeter on a three-quarter-mile gravel path. The shoreline features high sandstone bluffs. The island is named for Governor Leonard Farwell who in 1858 donated the land for a state hospital.[72]

North of Lake Mendota on Sherman Avenue is the North Unit of Cherokee Marsh Conservation Park, the largest of three units—the North, South, and Mendota Units—all located on the east bank of the Yahara River upstream from its outlet into Lake Mendota. These marshes are preserved examples of wetlands that once lay over large areas of central Dane County, most of which have now been drained for farming or filled for urban development. They acted as natural spongy filters, taking runoff from surrounding higher lands, storing the nutrients to be used by their lush plant communities, and slowly releasing cleansed water to the lakes. The elimination of these broad wetland areas is part of the cause of the pollution that now plagues the Yahara Lakes.

Taken together, the three units of Cherokee Marsh make up the largest wetland in Dane County. They feature a total of 4.3 miles of hiking trails that

4.15 A view of Lake Mendota looking north from the UW–Madison campus. The tip of the peninsula seen extending into the lake is Picnic Point.

pass through lowland oak forests, marshland, aspen and willow groves, and shrub-carr, a wetland community containing tall shrubs such as red osier dogwood. Bird- and wildlife watchers, paddlers, students, and teachers have long used Cherokee Marsh for outdoor recreation and education.[73]

From Cherokee Marsh, get back on Highway 113 and go northwest a short distance to the junction with County Road M. Follow M southwest about two miles to Governor Nelson State Park on the northwest side of the lake. Hikers can enjoy the undeveloped lakeshore in the park and follow trails through oak savanna and prairie lands.

About four miles southwest of the park and just north of Highway M is the Pheasant Branch Conservancy, offering another good sampling of the lakeshore's native ecosystems, with parking areas on the east side of the conservancy on Valley Ridge Road and on the west side on Pheasant Branch Road. Hiking trails take visitors through oak savanna, a restored prairie, sedge meadows, mixed woodlands, and marshes. One overlook allows a remarkable view of freshwater springs boiling up out of deep sands deposited by the glacier. Other overlooks are found on Frederick's Hill, a high drumlin on which Indigenous people built mounds centuries ago.

Taking Allen Boulevard south from Pheasant Branch, merging onto University Avenue to go southeast and east, then turning north onto University Bay Drive will bring you to Picnic Point, the long narrow peninsula projecting from the south shore of the lake. This popular destination appears to be a sandspit. In fact, it is much more substantial—an eroded sandstone ridge that once separated the preglacial Yahara River, curling south then east across the basin, from an ancient tributary that flowed in from the west. Both streams flowed in deep valleys, and Picnic Point was a promontory standing 250 to 300 feet above their confluence. Both valleys have been mostly filled by glacial debris and glacial lake deposits. The tributary valley now lies deeply buried under University Bay, the large southwest quarter of the lake.[74]

Observatory Drive, just south of Picnic Point, offers a view of the south shore of Lake Mendota from a high vantage point. The drive connects the west end of the campus on University Bay to Memorial Union, at the heart of the campus. This drive over a large drumlin formed around 19,000 years ago allows for a splendid open view of Lake Mendota (Figure 4.15). Another way to see the south shore and University of Wisconsin campus is to walk or bike the lakeshore paths from Picnic Point east to UW–Madison's famous Memorial Union Terrace. About one city block west of the terrace, sitting right on the lakeshore, is the Hasler Laboratory of Limnology, the world-class facility where scientists work at the forefront of limnology in North America. The Union Terrace completes your loop tour of Lake Mendota.

Lakes Monona and Wingra

A good starting point on Lake Monona is Olbrich Botanical Gardens on Lake Monona's northeast shore. A lakeside beach and park lie on the shore, providing a picturesque view of central Madison (Figure 4.16), and across Monona Drive from the park are 16 acres of outdoor display gardens surrounding an indoor conservancy, which allows for restorative connections to the plant world throughout the year. From there, you can skirt the lake's north shore on city streets that allow access to a number of small lakeshore parks and beaches, including Yahara Place Park where the Yahara River flows into the lake near the northeast end of Madison's isthmus.

4.16 The view from Olbrich Park with the State Capitol and downtown Madison in the distance

A few blocks northwest of Yahara Place Park, take Williamson Street southwest about ten blocks to where it ends. There, bear left to go southwest along the lakeshore on John Nolen Drive, right on North Shore Drive to Proudfit Street, and then left on West Washington Avenue, which takes you to Brittingham Park, a large green space on the north shore of Monona Bay. Informative signs at the main parking area give you a clear sense of the lay of the land and lakes, as well as the history of how Monona Bay, an extension of the lake on its west end, was once joined with Lake Wingra to the west before a glacier separated them with a small moraine. From the park, a lakeside path runs along the east and south sides of the bay.

To get to Lake Wingra, head west from the bay on Drake Street or Erin Street, crossing the now-leveled moraine along which South Park Street runs. Both streets take you to Henry Vilas Park, a sprawling, picturesque green space on the northeast end of Lake Wingra. You can circumnavigate Lake Wingra almost completely and stay on public land, most of it undeveloped. From Vilas Park,

4.17 Big Springs in the UW Arboretum is one of hundreds of springs that feed Lake Wingra.

Edgewood Park and Pleasure Drive (not open to cars) take you southwest for almost a mile through a lakeside lowland forest bordering a large cattail marsh on the lakeshore. The next green space, Wingra Park, accessible via Monroe Street, features a broad open space, a boat landing, and a lagoon nestled into a shoreline woods.

Wingra Park is adjacent to the UW–Madison Arboretum, which surrounds the rest of the lake—its southwest end and south and east shores. The Arboretum is a treasure left by the glacier, preserving a vast space on the southwest side of Madison in much the same state as it was prior to European arrival (Figure 4.17). It straddles a segment of the west side of the Yahara River Valley, its northeast side lying on the bed of Glacial Lake Yahara and its southwest side sitting higher, on a small recessional moraine.[75] In addition to much of Lake Wingra and its surrounding wetlands, the Arboretum hosts native woodlands, savannas, and Curtis Prairie—the world's oldest ecologically restored prairie. Leaders of this prairie's restoration, which began in 1933 in an abandoned horse pasture, were pioneers in the field of restoration ecology, including John Curtis, Norman Fassett, John Thompson, and Aldo Leopold.[76]

A combination of city streets, a bike path, and Wingra Creek all connect Lake Wingra to Lake Monona. A good place to wind up the tour of Lakes Wingra and Monona is at Olin-Turville Park on the southwest shore of Lake Monona. It hosts an open green space and a wooded area, all situated on a bluff overlooking the lake with trails to the shore. Several vantage points give visitors a view of the entire length of Lake Monona, looking across to the northeast shore where Olbrich Gardens is situated. This bluff over the lake is one of the points from which you could look down into the deep preglacial Yahara River Valley if you could wander back in time.

Lakes Waubesa and Kegonsa

The Yahara River leaves Lake Monona from a bay on its south side and angles southeast through a vast lowland designated as Capital Springs State Recreation Area, which hugs the north end of Lake Waubesa. This lowland was part of a long bay of Glacial Lake Yahara that stretched west-southwest for four miles to where Fish Hatchery Road now runs south.[77] The area features several large springs feeding a diverse landscape that includes wetlands, prairie, woodlands, and effigy mound sites, accessible on more than six miles of hiking trails. Access Capital Springs by taking South Towne Drive off US Highway 12/18 (West Beltline) south for a mile and turning left (east) on Moorland Road, which becomes Lake Farm Road.

In Capital Springs, the Yahara River flows through little Upper Mud Lake and into Lake Waubesa. Boaters and anglers love Waubesa, which is oblong and arcing, looking something like a fish itself, as depicted on a map. Babcock County Park, located halfway down the east shore of Lake Waubesa in the village of McFarland, has a busy boat landing and campground. Here is where the Yahara River exits the lake to meander southeast. On the southwest end of the Lake Waubesa is the Waubesa Wetlands State Recreation and Wildlife Area (Figure 4.14). Here again, wetlands have gradually taken over a bay of the shallow lake that once extended farther southwest.

This spring-fed wetland complex—a mix of sedge meadow, fen, and shrub-carr communities lying over peat deposits up to 95 feet deep—is one of the most diverse and pristine in southern Wisconsin. The Wisconsin Department of Natural Resources website description makes it sound positively otherworldly,

4.18 The Yahara River exits the basin of Lake Kegonsa through this broad wetland.

with "quaking sedge mats, calcerous fens . . . deep-spring cones lined with filamentous algae and purple-colored bacteria, . . . grass-of-parnassus, . . . American woolly-fruited sedge, blue-joint grass . . . and sage willow." This 566-acre aquatic cornucopia provides a habitat for sandhill cranes, green herons, marsh and sedge wrens, blue-winged teal, green-winged teal, and willow flycatchers.[78]

The southernmost lake in the Yahara chain is Lake Kegonsa. From McFarland, US Highway 51 runs southeast for three miles to County Road AB. Take this road as it arcs northeast along the lakeshore for a mile to Fish Camp County Park on Lone Tree Point, where the Yahara River flows into the lake. True to its name, the park is on the site of an old state-run fish camp. From 1932 to 1969, crews toiled year-round on this and other lakes to remove exploding populations of invasive carp that were damaging lake ecosystems. Carp had been introduced across Wisconsin around 50 years earlier by the Wisconsin Fish Commission, seeking to create a new food source for the state's growing population, but as it did in many other places, the plan backfired. Signs in the park tell the story of the grueling life of fish camp workers.[79]

A short distance along the shore, on the northeast corner of the lake, is Lake Kegonsa State Park, accessible via County Roads AB and MN and Door Creek Road. Its five miles of hiking trails allow exploration of a variety of lakeshore, woodland, wetland, and restored prairie ecosystems. Bird-watchers have spotted more than 60 species of birds here, including pheasants, ducks, grebes, horned larks, kingbirds, and increasingly rare bobolinks.[80] Most of the park was once on the bottom of Glacial Lake Yahara. The higher areas on the park's south side, known as "the oak knolls," were islands in that lake.[81]

Next to the state park on the east shore of the lake is LaFollette County Park, adjacent to a vast, lush wetland (Figure 4.18) where the Yahara River leaves its last lake basin and enters its final stretch. Along the way, it passes through the thriving city of Stoughton, famous for touting and fiercely defending its deep Norwegian cultural roots. People from there have referred to Madison as "the town on the north side of Stoughton." From there, the Yahara River takes a meandering course of 15 miles to where it joins the Rock River.

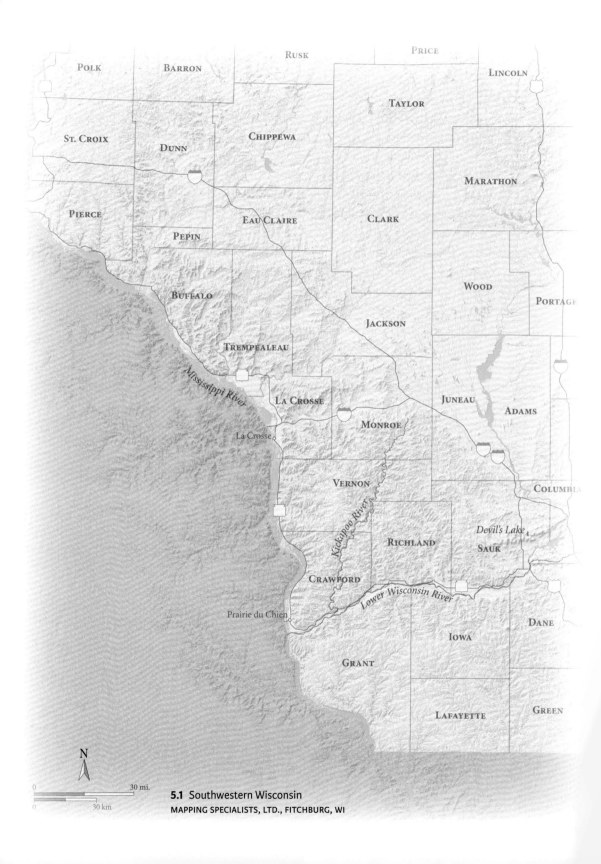

5.1 Southwestern Wisconsin
MAPPING SPECIALISTS, LTD., FITCHBURG, WI

5

THE DRIFTLESS AREA

Unlike the surrounding northern region of North America, southwestern Wisconsin—the Driftless Area—was never made over by glaciers, so the landscape is among the oldest in the world. The story of this area's water is mostly one of flowing water as an agent of erosion working uninterrupted for hundreds of millions of years.

The Driftless Area was once a plateau formed on the floors of ancient seas. After the last of the seas had departed the region, the plateau was thickly covered by layers of dolomite overlying Cambrian sandstone. These rock layers had buried the Baraboo Hills, the Blue Mounds, and all the other promontories that now stand high in the Driftless Area. Since the withdrawal of the last ancient sea, rain, flowing water, wind, and frost action have carved the ancient plateau deeply. When rainwater first fell on the plateau, it pooled and flowed, taking the paths of least resistance. It formed rivulets that joined to make larger streams. These developing streams gradually etched the water-soluble dolomite and formed shallow channels. As more water flowed, the streamlets merged to form creeks and rivers that carved through the dolomite and deeply into the softer underlying sandstone.

Over hundreds of millions of years, such quickening erosion formed the dendritic, or branching, pattern of stream valleys that decorate the Driftless Area today. Locally, these narrow, steep-sided valleys are called coulees—derived from the French word *couler*, meaning "to flow." They are separated by high ridges that vary in width from a few feet to several miles. The longest, widest, and highest of the ridges, called Military Ridge, lies along a line running from western Dane County all the way to the Mississippi River. It is named for the road that was built in 1835 to connect Fort Crawford in Prairie du Chien with Fort Howard in Green Bay.

Military Ridge forms a divide between two watersheds. On its south side, streams flow south and southwest across a long, gentle slope, emptying into rivers that flow to the Mississippi. On its north side, streams tumble down the much steeper slope to the Lower Wisconsin River. Although that river eventually became the main trunk in the Driftless Area's arterial system of streams, the broad Wisconsin River Valley once looked more like the rest of the Driftless Area—ragged and narrow—until after the departure of the glaciers, which imparted great floods that widened and deepened the valley.

Because of the long-term pattern of erosion, there are no sizeable lakes in southwestern Wisconsin, although a number of dams have been built to create small lakes. However, in the absence of glaciers, streams have ruled in the Driftless Area, continually carving their valleys deeper and longer and draining away any water that would have accumulated naturally on the land.

DEVIL'S LAKE

Devil's Lake is at the heart of Wisconsin's most popular state park, receiving 1.5 million visitors per year, on average. It's probably safe to assume that almost every one of the park's visitors has uttered some expression of awe or gazed a little slack-jawed at the soaring palisades flanking the lake. In fact, the gorge has surely awed its human visitors for as long as humans have lived in or near it. For centuries, Native Americans who lived in or near the gorge regarded it as a sacred place and gave the lake names that translate variously to Spirit Lake, Mystery Lake, Holy Lake, and Lake of the Red Mountain Shadows.[1] For those people and their descendants, it has been a place for hunting and fishing, but also a place for wonder and worship of the spirits. Something similar could be said of many, if not most, of today's visitors to Devil's Lake State Park.

The lake itself could be called ordinary—a roughly rectangular body of water, 1.3 miles long, half a mile wide, and at most 45 feet deep. It is the gorge that frames the lake that makes it a special place. The cliffs that rise 500 feet above the east and west sides of the lake are made of a reddish-purplish rock called Baraboo quartzite, which is among the hardest types of rock on the planet. It was forged from ancient sandstone between 1.65 and 1.45 billion years ago when the primitive North American continent, driven by tectonic forces, collided with a smaller continent to the south. In the coastal collision zone, land was crumpled and folded in complex ways, and the mass of sandstone beneath was churned and metamorphosed by heat and pressure. One area of this crumpled land is now called the Baraboo Hills, for which the area's unique type of quartzite—stained red and purple by iron oxides—is named.[2]

The end result of the continental collision was a low mountain range stretching from southern Wisconsin to the west, possibly as far as the Dakotas. The Baraboo Hills are a remnant of those mountains, left standing after the rest of the range was eroded away or buried under the sediments of invading seas. This unique arrangement of bluffs is shaped something like a rowboat pointed east, with the boat's gunnels and stern forming three ranges of hills, and the boat's interior forming a central lowland. During the tectonic collision, the north and south ranges of hills—the two sides of the boat—were heaved up like folds in a rug. Between the two ranges, the land was depressed into a trough formation called a syncline.[3]

Before the Cambrian sea spread over the region more than 500 million years ago, the quartzite hills stood an estimated 1,000 feet above the surrounding barren, rocky plain. For 100 million or more years, Cambrian and later Paleozoic seas advanced and retreated multiple times. For much of that era, the bluffs of the future Devil's Lake gorge stood as islands, their cliffs often slammed by tropical storm waves. Eventually, the invading seas filled the valleys among the Baraboo Hills with sedimentary rock and, finally, buried the entire range.[4]

When the last of the seas finally retreated, erosive forces went to work on the sedimentary rock layers that mantled much of Wisconsin. In the Devil's Lake area, a network of streams flowed and carved shallow lowlands into these layers of Paleozoic sandstone and dolomite. Over millions of years, the stream valleys deepened and, eventually, the quartzite bluffs emerged once again, standing up to the wind and water that had carried the overlying sedimentary rock away. By the time the first glacier arrived, ancient rivers had bored out the 800-foot-deep gorge that would later hold Devil's Lake. The gorge of those days was 300 to 400 feet deeper than it is today.

Geologists are quite sure that no glaciers advanced into the Devil's Lake gorge because certain complex rock features there would have been toppled by a wall of ice pushing through. Thus, the western third of Devil's Lake State Park, including the lake gorge, is in the Driftless Area. The most recent glacier moved in from the east about 19,000 years ago. It crept up the easternmost slopes of the

5.2 A view from the south shore of Devil's Lake

Baraboo Hills, and in the southern range, it stopped in the highland just east of the ancient river valley where the Devil's Lake gorge would be formed. North and south of that highland, the ice kept pushing westward into the valley, but there it stopped, plugging the valley in two places that would become the north and south ends of the gorge. Those walls of ice were possibly 200 feet high.

Glacial ice did not pulverize quartzite as it did some other types of rock, but the glaciers had an indirect effect on the quartzite hills—more like that of a pickaxe than a bulldozer. Even in glacial times, during the short summer months, water would flow into cracks and crevices in the quartzite cliffs. During the long, frigid glacial winters, ice formed and expanded within those cracks and crevices. With each freeze-thaw cycle, the ice widened the cracks, and over time, large chunks of quartzite were pried away from the cliffs and tumbled to the ground

5.3 The iconic Devil's Doorway at Devil's Lake

below. One result of this process is the massive jumble of quartzite blocks, some larger than 10 feet on a side, draped along the bases of the cliffs around the lake in what are called talus slopes (Figure 5.2).

Likewise, the glacial freeze-thaw cycle is responsible for some of the most popular rock features in the lake gorge, including the seemingly precarious Devil's Doorway (Figure 5.3). Piece by piece, over many centuries, chunks of quartzite from all around this structure must have either fractured and fallen away or been knocked down by other quartzite blocks falling from above. In this way, the freeze-thaw pickaxe effect literally sculpted this and other amazing rock features in the gorge.

Around 15,000 years ago, the climate began to warm, and as the ice walls at the ends of the gorge began to melt, water pooled between the two walls, forming a glacial lake—the original version of the lake we know today. Its waters rose to somewhere between 90 and 150 feet above the present-day lake level, making it three to four times deeper than modern Devil's Lake. Then, the lake's waters found an outlet at the northwest corner of the basin and flowed into vast Glacial Lake Wisconsin to the north. Over time, icebergs calved from the glacier's walls and plunged into the lake. We know this because glacial erratic boulders have been found in the small valley at the southwest corner of the lake where Messenger Creek now flows in. They must have been carried there by icebergs because this location was never glaciated. Today, divers also find such erratics on the lake bottom where they likely fell when their ice rafts melted away.

The glacier left two high ridges spanning the gorge at either end of the lake. They are actually connected as part of the Johnstown Moraine, which snakes through the highlands east of the lake and stretches away from the gorge to the north and south. The Devil's Lake State Park Nature Center sits near the crest of the north moraine ridge. The south ridge overlooks Roznos Meadow, a broad plain to the east from which the moraine rises abruptly to 150 feet above the meadow. Meltwaters rushing off the walls of ice on both ends of the gorge washed huge loads of sand and gravel into the lake's basin, filling it about halfway. Geologists estimate that this outwash now on the lake bottom is 300 to 350 feet deep.[5]

As the glacier finally melted back and meltwaters dwindled, the new lake began to seep through its sandy, gravelly bottom into the groundwater system, and the lake level eventually stabilized at roughly its current level. It is fed by one inlet stream, Messenger Creek; by the rainwater and snowmelt collected by the gorge; and by springs on the lake bottom. Devil's Lake has no outlet streams. Because the lake is perched on a thick mass of sand and gravel, placing it higher than other nearby lakes and streams, water seeps out of the lake bottom, finding its way through groundwater to other waterways and thereby keeping the lake at a fairly predictable level.

Devil's Lake is now preserved in a relatively pristine condition within the state park, but it was not always so. In the 1800s, loggers leveled much of the surrounding forestland, although a few forested areas in the park were spared and are today considered old-growth stands.[6] A railroad was extended into the gorge in 1872, and from that time on, a tourist trade flourished. By 1900, Devil's Lake was the site of four hotels, and cottages sprouted along its shores. At some point, nine passenger trains arrived at the lake's railroad station every day, and excursion boats circled the lake. However, this flurry of commercial tourism came to an end within three decades, as a spirit of preservation drove an effort to protect the lake gorge. Devil's Lake State Park was established in 1911.

Devil's Lake is treasured for many reasons. Its beaches provide a cool respite for both young and old on hot summer days. Divers enjoy its cold, clear waters. Rock climbers tackle the roughly 2,000 routes among the cracks and overhangs on the gorge's walls. Students of all ages and their instructors arrive to study, teach, and learn. Scientists from around the world probe its forests and glades. And from the trails along the bluffs, hikers gaze down onto the treetops, their colors changing with the seasons, on the forested moraines that bracket the lake gorge. Watchful visitors can see gangly yet graceful herons gliding between their rookeries and the lakeshores, vultures soaring over the cliffs in search of prey, or any of nearly 100 other species of birds.

The lake and its environs—the Baraboo Hills—are also treasured as repositories for an ever-dwindling number of plant and animal species, many of them rare, threatened, or endangered, most of them having lived in the hills for thousands of years. Former Devil's Lake State Park naturalist Kenneth Lange has described the Baraboo Hills as "an island of relatively unfragmented forest in a sea of agricultural land and woodlots." In all, the hills have more than 900 species of vascular plants, close to half the number of such species in Wisconsin. Also, almost a third of the state's nonnative (introduced) plant species have populations in the hills. In terms of animal species, the Baraboo Hills have populations representing 60 percent of all the state's fish species, 89 percent of amphibians, 56 percent of reptiles, 59 percent of birds, and 67 percent of mammals.[7]

The lives of these plants and animals are intertwined in communities, some quite fragile, sheltered in the hills. Around Devil's Lake, two increasingly rare community types are rock glades and cliff communities, where some threatened and endangered species live. Mammals under threat or of special concern in the Baraboo Hills include the big brown bat, long-eared bat, Franklin's ground squirrel, and western harvest mouse. Birds under threat or of special concern include the Bell's vireo, Cerulean warbler, red-shouldered hawk, trumpeter swan, northern bobwhite, and osprey.[8] Devil's Lake and the Baraboo Hills are a storehouse of biodiversity, making them precious not only to people, but to the entire biosphere in southern Wisconsin.

TRAVEL GUIDE
Devil's Lake

An excellent way to explore Devil's Lake is by hiking all the way around it on four trails—the East Bluff, CCC, Grottos, and West Bluff Trails. Start at the East

Bluff trailhead on the north side of the lake out of a parking area due east of the park headquarters. It is a moderately difficult trail with some steep sections; however, much of it is paved with asphalt and includes flights of stone steps.

On the East Bluff Trail, after a steep first segment, the trail levels off and turns sharply right, continuing up toward the bluff top. To the left of this turn is Elephant Cave—a shallow cave tucked in under massive blocks of sandstone sitting on quartzite. A spur trail that branches to the left past the cave—an unmarked, dead-end trail—leads to an excellent example of a conglomerate wall made of quartzite stones and Cambrian sandstone layered in with larger quartzite blocks. Geologists interpret this layered wall as evidence of an ancient cliffside on an island in the Cambrian sea. The main cliff was near here, and fragments of it were likely loosened and dropped by powerful waves lashing the shore and then rounded by centuries of wave action. Later, they were buried by other sediments and compressed into conglomerate as the sea level rose and more layers were deposited.[9]

On this trail, hikers can also see ripple marks frozen into flat rock surfaces—another indication that an ocean shore once existed in the area. On that ancient shore, waves rippled the sand just as they do today on the beaches of Devil's Lake. As the seas quietly advanced and coastal waters deepened, these ripple marks were buried by muck and the remains of animals and plants that later became layers of sedimentary rock. During metamorphosis, such interlayered ripples were preserved in quartzite. Look for them on the flat sides of boulders on the talus slopes and on some of the flat stones that now serve as steps on the trails.

Back on the main trail at Elephant Cave, it is a short climb past the cave to the first view of the lake from the top of the bluff. As you venture onto the bluff top, remember that you stand atop a mass of quartzite that is 4,000 feet deep and about 1.6 billion years old. From here, you'll begin a rocky up-and-down traverse on a trail across the top of the bluff. At several points on this part of the trail, you can walk to the edge of the bluff for views of the north beach area and the moraine behind it. There is where the 200-foot wall of ice stood between 15,000 and 12,000 years ago, dropping the gravel and boulders that now underlie the forested ridge. The city of Baraboo is also visible to the north.

Closer to the south end of the lake, the East Bluff Trail veers to the east at its junction with the Balanced Rock Trail, on which you can descend to the valley floor. This is a steep, difficult trail, one of three that descend across the

5.4 Balanced Rock, a favorite destination at Devil's Lake

talus slope. However, it affords superb views of the gorge and leads you to the famous Balanced Rock (Figure 5.4), sculpted from quartzite by countless cycles of freezing and thawing during glacial times.

Beyond its junction with Balanced Rock Trail, East Bluff Trail continues along the rim of the gorge with spectacular views of the forested moraine rising away from the flat south beach area. Another wall of ice stood here 15,000 years ago. If you could take yourself back to that time and wait long enough, you might hear and see a block of quartzite peeling away from the rim of the gorge and crashing down onto the talus pile in front of that wall of ice. Shortly beyond the point where the trail turns east, the Devil's Doorway Trail splits off to the right. This is a short loop trail that takes you to the iconic rock feature that is probably the most photographed of all the park's attractions (Figure 5.3). It is a very steep trail with high stone steps to manage, but it is only a tenth of a mile long and affords superb views of the south gorge and quartzite features.

When you gaze on seemingly precarious formations like Devil's Doorway and Balanced Rock, you might wonder what it would take to tip them over. The answer is: more than we can easily imagine, because quartzite is an extremely dense and heavy rock. Just one cubic foot of quartzite weighs 165 pounds. With that information, and considering that Balanced Rock is approximately 12 feet tall, you could take a stab at estimating its weight.

Farther along on the East Bluff Trail, a second route to the valley floor splits off to the right. This is the Potholes Trail, another steep hike across the talus slope, but you need only descend 40 or 50 steps to see the depressions for which the trail is named—potholes that were carved into the quartzite by powerful eddies in an ancient river that whirled rocks and gravel against the streambed hundreds of millions of years ago. They are near the junction with the East Bluff Trail, on the west side of the Potholes Trail, and they indicate that powerful erosive streams once flowed at this level, high over the present-day lake.

Beyond the Potholes Trail junction, it is another half mile on the East Bluff Trail to the CCC Trail, the third and last trail down to the valley floor. From this junction, you can continue eastward on a trail along the bluff, but the East Bluff Trail ends here. What continues east is part of the Devil's Lake Segment of the Ice Age National Scenic Trail, which eventually descends the bluff and crosses the south moraine, dropping into Roznos Meadow. Of the three talus slope descents, the CCC Trail, named for the Civilian Conservation Corps that created it in the 1930s, provides the best unblocked views of the south gorge, along with good closeup looks at the talus. It also passes a few flat rocks on which to sit, rest, and contemplate the vast gorge and all that has happened here.

The CCC Trail descends to meet the Grottos Trail, which runs west along the base of the talus field. An easy, wide trail, it follows a gentle downslope across outwash from the south gorge glacial wall. From this trail, you can step down into grottos—low spaces off the trail, walled by boulders at the bottom edge of the talus slope. There, you get a stirring view of the talus, looking up at the enormous rock pile and the cliffs 500 feet above you, from which these boulders have fallen for thousands of years.

At the west end of the Grottos Trail is a well-traveled path across railroad tracks to the south shore parking area. At this point, you have merged with the Devil's Lake segment of the Ice Age National Scenic Trail. Follow it across

5.5 The East Bluff at Devil's Lake viewed from the West Bluff Trail

the beach area to the boardwalk on the south shore of the lake. From there, it is about a mile to the crossing over Messenger Creek and, shortly beyond, the junction with the West Bluff Trail, a one-mile trail comparable to the East Bluff Trail in difficulty. The first stretch of about half a mile is a steep climb on blacktop through the woods to the first spur trail that leads to the bluff edge, where the views of the lake and the east bluff are superb (Figure 5.5).

From this point, note as you look at the East Bluff that the vegetation pattern slopes down about 20 degrees from south to north (right to left). This shows that the rock layers of the bluff are tilted, due to the crumpling of the land described in the beginning of this chapter. The bluff you see is literally a cross section of the south range of the Baraboo Hills that shows how the rock layers are tilted. From the West Bluff, you can also clearly see how both ends of the gorge are blocked by moraines where walls of ice once stood. In Ice Age days, you would have seen the East Bluff between the ice walls with no ice on top, but no trees either. Lichens and a few other primitive species were the only forms of life on the bluffs in those days.

The West Bluff Trail continues with some steep ups and downs, then finally skirts the north end of the bluff and drops down across the north moraine into the north beach area. At this point, you have completed a grand loop that gives you a sampling of all the major geologic features of Devil's Lake and its magnificent bluffs.

THE KICKAPOO RIVER

The Kickapoo River Valley has been called the heart of the Driftless Area. The river itself might be called the aorta of the Driftless Area. Except for the Wisconsin River, the Kickapoo is the largest of the many streams and rivers that sustain life there and make the area unique. It was flowing water that carved up the Ordovician plateau over hundreds of millions of years, creating an array of high ridges separated by a branching, or dendritic, pattern of steep-sided valleys—classic coulee country never reshaped by glaciers. Just as the Driftless Area is one of the oldest landscapes in the world, the Kickapoo and its many tributary arteries make up one of the world's oldest river systems.

The Kickapoo River is famous for its meandering nature, flowing 125 miles within a watershed half as long. Its name, given by Algonquian-speaking people, means "he moves about, standing now here, now there," a fitting description for a river that nearly meets itself in countless places along its route. The river originates in two valleys that merge north of the Driftless Area's geographic center, just northwest of the village of Wilton. It drops just five feet per mile, on average, in its ever-winding southwesterly journey from there to the village of Wauzeka, where it empties into the Wisconsin River.[10]

The Kickapoo flows within a deep bed of sandstone laid down more than 500 million years ago. Over time, succeeding ancient seas deposited more layers of sandstone, shale, and, dolomite. Layers of these various rock types are clearly exposed in bluffs along the Kickapoo, some rising higher than 400 feet above the river.

Because some of these layers are softer and more erodible than others, the bluffs show off exquisite rock sculptures carved by wind, flowing water, and the winter frost cycle over millions of years. In many places where the harder layers of rock form ledges and the softer sandstone beneath them has been eroded out, rock overhangs have emerged. These provided shelter for people living in the

5.6 The Kickapoo River Valley viewed from the Hemlock Nature Trail in Wildcat Mountain State Park. GAIL MARTINELLI

area hundreds to thousands of years ago. They are also the sites of spectacular ice formations in winter months. Water seeping from cracks and crevices in the rock throughout the winter forms ice floes across the openings of these rock shelters, often referred to as ice caves.

As with its topography, the story of the Driftless Area's flora and fauna is different from that of the rest of the state. The area is thought to have hosted a limited diversity of plants and animals throughout the Ice Age, earning it the ecological designation "refugium landscape"—an area where special circumstances have allowed for the survival of natural communities that have become extinct on surrounding lands. Lichens, hardy plant species, small mammals such as chipmunks, and possibly some amphibians and reptiles survived the frigid glacial conditions by hunkering down in the deepest valleys of the Driftless Area. After the ice retreated for the last time, and as the climate warmed, these populations slowly grew and spread across the state. Thus, it's possible that the bulk of the state's plant cover before 1800 derived from Driftless Area stock.[11]

The postglacial plant cover in the Driftless Area likely resembled that of the time before glaciers invaded. On the ridgetops a mix of prairie and oak savanna had evolved, grading into thicker oak woodlands on the flanks of the highlands. Mixed forests had become established in the deeper parts of the valleys where moist, cool conditions prevailed.[12]

In the broad Kickapoo River watershed, mixed hardwood forests dominated, composed of sugar maple, hemlock, beech, yellow birch, basswood, elm, and ash. The Kickapoo Valley was surrounded by the less diverse but more typical Driftless Area mosaic of prairie, oak savanna, and oak woodlands.[13] Perhaps the valley, being near the center of the Driftless Area, was also at the heart of the refugium landscape where plant life survived the Ice Age and served as a source of spreading plant communities after the glaciers were gone.

The Kickapoo Valley has been occupied by people on and off for thousands of years, according to evidence gathered from more than 450 archeological sites in the upper valley alone. Evidence dating between 12,000 and 850 years ago—spanning the Paleo through Woodland traditions—includes rock shelters, camp and village sites, burial mounds, and petroglyphs (rock carvings).[14] Artifacts from the late Paleo people, between 8,000 and 10,000 years old, include projectile points and stone adzes.[15]

Woodland Indians who were in the area as long as 2,500 years ago apparently used the valley as a seasonal hunting ground as they traveled between more permanent settlements northeast of the valley and the mouth of the Kickapoo on the Wisconsin River. At some point, a small band called the Ocoche lived in the valley; a group of high bluffs in the valley called the Ocooch Mountains were named after this tribe.[16] When Europeans first arrived in the valley, it was occupied by Sauk, Meskwaki, and Ho-Chunk people who, in addition to hunting and farming in the area, were making use of lead they dug from shallow pits in Wisconsin's future lead mining district, located mostly south of the Kickapoo Valley. According to some sources, another group called the Kickapoo lived in the river valley before 1700.[17]

Today, large areas of the Kickapoo Valley look much as they did in presettlement times. Thick stands of mixed hardwood forest and even some smaller stands of old-growth hemlock forest still shelter the quiet floodplain where the river meanders as it has for thousands of years. Increasingly rare cliff-clinging

5.7 One of many Cambrian sandstone bluffs along the Kickapoo River that hosts ancient plant communities

plant communities still grace the ancient sculpted bluffs that stand along some stretches of the river (Figure 5.7). However, the Driftless Area surrounding the valley has changed from a postglacial mosaic of prairie, oak savanna, and oak woodlands to pockets of mixed hardwoods scattered among farmland dotted with small towns. In all of southern Wisconsin, prairie and savanna have nearly vanished; oak savanna now occupies less than one hundredth of a percent of the range it occupied prior to European immigration.[18]

Changes in the Kickapoo Valley would have been far more dramatic had the federal government carried out its plans for damming the valley. In the 1960s, several communities in the lower Kickapoo Valley experienced serious flooding, and the proposed solution was to dam the river in the upper valley near the village of La Farge. Some 149 families sold or were forced off their land to make way for this project. However, cost overruns, concerns about the likely degraded water quality in the lake-to-be, and loud protests by some valley residents and environmental groups put a halt to the project in 1975. Meanwhile, downstream of the proposed dam, several communities did work to prevent or mitigate flooding in their towns.[19]

Part of the valley that would have been flooded lies within Wildcat Mountain State Park, established in 1948. The rest of it is now preserved within the 8,600-acre Kickapoo Valley Reserve, controlled jointly by the Ho-Chunk Nation and the state of Wisconsin since 2000, when the federal government transferred ownership of the area to those parties. Within the state park and reserve, canoeists now enjoy a stretch of the river that runs past picturesque, ancient bluffs of sandstone (Figure 5.7). These areas are contiguous and subject to state and tribal laws, which now protect all of the Upper Kickapoo Valley from disruptive development. The Kickapoo Valley Reserve has an impressive, highly informative visitor center in La Farge and has been designated a National Natural Landmark.

The story of flora and fauna in the Kickapoo Valley is a story of widespread decline, both of ecosystems and species, largely due to European immigration, settlement, and farming. But it is also a story of the recovery of many systems and species since preservation of the valley became a priority. When European farmers came, they drained wetland areas and cleared grasslands to make way for their crops. At one point, farms occupied more than half of the Kickapoo River system's watershed.[20]

Prior to settlement, the wetlands and grasslands had acted as sponges that absorbed precipitation and fed it slowly to streams and rivers. With the conversion of these areas to cropland, away went that natural flood control service. Soon, summer rainstorms and spring snowmelts resulted in soil-laden water surging off the land and into the stream valleys and towns that had grown up there. Meanwhile, with the disappearance of aquatic plants and prairie grasses from the cleared areas, the animals that had lived there evacuated or died off. The food webs stitched together over millennia were destroyed—replaced by monoculture crops with the soils eroding from under them. By 2000, eroded soil lay several yards deep in some of the Driftless Area's stream valleys.[21]

Much of this degradation of natural systems has now been reversed, thanks to preservation of the Upper Kickapoo Valley. Wetlands and grasslands are now reclaiming some of the farmland abandoned during the failed 1960s dam project. Flooding is still a problem in the valley, but it would be worse if these natural systems were not being restored. Also, in the 1920s, after nearly 75 percent of the Driftless Area's farms were found to be degraded by soil erosion, the federal government in collaboration with farmers began successful experimental soil conservation programs that have since benefited the Driftless Area and the whole country.[22] It is interesting to note that the soil conservation practices applied consistently by Driftless Area farmers could do as much to control flooding as the damming of the Kickapoo would have done—at a fraction of the cost and with none of the environmental damage or human displacement that a dam would have caused.

The Kickapoo Valley's forests and wetlands host rich biodiversity. As in postglacial days, the deepest forests are made up of varying combinations of oak, maple, hickory, pine, and hemlock, with minor stands of cedar and tamarack. In the Kickapoo Valley Reserve, wetlands occupy 17 percent of the acreage—a considerable improvement over the time of peak agriculture in the valley. Protected forests in the lower valley are among the largest, most intact forest stands in the Driftless Area.[23] Prairie areas are being restored jointly by the Ho-Chunk, the education-focused group Prairie Enthusiasts, and the Wisconsin Department of Natural Resources. The valley's 400 or more plant species include at least 12 rare or endangered species, notably rock club moss, cliff cudweed, northern monkshood, and Lapland azalea.[24]

The Kickapoo Valley shelters a rich assortment of animals. The wetlands and river banks are frequented by beavers, otters, and muskrats, and the forests are home to mink, raccoons, woodchucks, foxes, fox squirrels, and white-tailed deer. More than 100 species of birds in the valley include rare species such as red-shouldered hawks; Acadian flycatchers; cerulean, Kentucky, worm-eating, and blue-winged warblers; and Louisiana waterthrush. Other birds commonly sighted include bald eagles, sandhill cranes, green herons, wood thrush, Bell's vireo, brown thrashers, bobolinks, and eastern meadowlarks. The valley's reptile populations include painted turtles, western fox snakes, northern redbelly snakes, and the threatened Blanding's and wood turtles. Amphibians include bull, green, leopard, and wood frogs; chorus and spring peeper tree frogs; and blue-spotted and four-toed salamanders. Fish populations living in the Kickapoo and its tributaries include brown, brook, and rainbow trout, making these streams highly valued among anglers.[25]

The Kickapoo Valley has endured many disturbances, both natural and human-made, and at this point, it is recovering impressively from these degradations. However, like most natural settings, it remains vulnerable to abuse and disturbance. Perhaps the best advice we can take when visiting the valley is summarized in these words from the Kickapoo Valley Reserve Visitor Guide, provided by a Ho-Chunk writer whose ancestors lived in the valley for centuries: "As you walk in the Reserve, you are part of the land and its spirit. Be humble in its presence and respect its power to change, provide, and continue."[26]

TRAVEL GUIDE
The Kickapoo River

The Kickapoo is best seen by taking a slow, meandering canoe trip, but the valley can also be explored by driving, mostly along State Highway 131 from Wilton to Wauzeka. Wilton lies on the river and also on the popular Elroy-Sparta State Trail—the first rail-to-trail conversion in the country—which passes through the headwaters region of the Kickapoo Valley, with views of wetlands that feed the river.

The next town downstream from Wilton is Ontario, where State Highway 33 splits off from 131 and leads to Wildcat Mountain State Park, a must-see destination for anyone visiting the Kickapoo Valley. Here, you can get a good feel for what Wisconsin was like before the glaciers gave the land a makeover, with rocky ridges rising abruptly from valley floors and steep-sided, forested hollows etched into the ridges. The Hemlock Nature Trail gives the best sampling of the park's terrain. This 1.3-mile loop trail to the top of Mount Pisgah is somewhat rugged and difficult with a total elevation change of 365 feet, but it is well worth the effort.

The west side of the loop passes through majestic stands of white pines and virgin hemlock overlooking the river (Figure 5.6)—rare old-growth trees, for this part of the park was never logged. The sandstone underfoot is porous and holds water, which helps to keep the slopes cool and moist and suitable for these trees. The trail meanders away from and back toward the river several times as it crosses the little valleys cut into the land by seasonal tributaries feeding the Kickapoo. On this trail, hikers climb through time—many thousands of years for every foot of elevation—as they rise among the layers of Cambrian sandstone. The east side of the loop trail passes spectacular sandstone outcroppings, intricately eroded by wind, water, and ice (Figure 5.8).

Because this area is unique and pristine, it has been protected as a state natural area. It hosts a number of rare plant species, some of which were here before the glaciers, including the low-growing flowering plants Sullivant's coolwort and moschatel.[27] Look, also, for walking fern, whose long arching leaves radiate

5.8 This sandstone outcropping in Wildcat Mountain State Park was carved mostly by flowing water.

from its roots. New plants sprout where its leaf tips touch down. It can be found "walking" across rock outcroppings. At the top of the ridge, an overlook allows a remarkable panoramic view of the Upper Kickapoo Valley.

Adjacent to the state park on its south side is the Kickapoo Valley Reserve (KVR). Crisscrossing the reserve is a large network of trails designated for hiking, mountain biking, horseback riding, and other uses. These trails provide an excellent sampling of Driftless terrain and of the verdant Upper Kickapoo Valley. The entire stretch of the river within the KVR is popular among paddlers for its views of the ancient bluffs and their fragile plant communities (Figure 5.7).

Old Highway 131 (which has been replaced by today's straighter highway) now serves as a hiking and biking trail. Its south end is close to the KVR Visitor Center, which makes a superb showcase for the KVR, its many volunteers, and the reverence they hold for the Kickapoo Valley. Completed in 2004, the center features interactive displays on the river's geology and natural history. The building was constructed partly of lumber from trees that were cut to make way for the rebuilt Highway 131, and it makes use of passive solar energy and a geothermal heating and cooling system.

Highway 131 continues south through the valley, passing through several small towns, each with its own unique identity. La Farge is known for the KVR Visitor Center and for its farmers' market, orchards, and CROPP Cooperative (which produces the Organic Valley brand), composed of more than 2,000 farmers dedicated to saving small family farms through organic farming. Soldier's Grove famously relocated its business district from the river floodplain to higher ground in the early 1980s after devastating flooding. In the process, the community built the country's first solar village. And picturesque Gays Mills is known for its annual Apple Fest and for community and cultural education opportunities such as the Kickapoo River Museum and Kickapoo Culinary Center.

Two miles south of Gays Mills is the Kickapoo River State Wildlife Area–Bell Center Unit, with one of the largest intact forests in the Driftless Area. It provides vital habitat for a number of forest interior bird species that are declining, notably the red-shouldered hawk, Acadian flycatcher, wood thrush, and Kentucky and cerulean warblers. There are no designated trails, but hiking is allowed and accessible from two parking areas in the village of Bell Center.[28]

Ten miles south of Bell Center on Highway 131 is the village of Steuben where State Highway 179 goes west. Just over a mile from this junction, turn onto Kickapoo Valley Road heading south along the river. This is a narrow gravel road that requires a slow, careful passage, but it affords delightful views of the Lower Kickapoo Valley where the river meanders lazily. It flows among countless little ponds, oxbows, and backwaters nestled into the saturated broad floodplain, replenished annually by spring flooding. Some stretches on this road can feel positively otherworldly, even a little spooky, giving one a strong sense of having traveled back in time.

Kickapoo Valley Road rambles for about six miles to its junction with Highway 131 just north of the Kickapoo River State Wildlife Area–Wauzeka Unit. Like the Bell Center Unit, it protects a large swath of habitat for a rich diversity of birds and other wildlife. It has no designated hiking trails but is accessible from a parking area on Highway 131 shortly north of the village of Wauzeka.[29] Highway 131 ends at State Highway 60, which crosses the Kickapoo near this junction. The river then flows on for a few miles to empty into the Wisconsin River.

For an alternate view of the Lower Kickapoo Valley, take Highway 60 west into Wauzeka, and then drive north on County Road N one mile. Here, Plum Creek Road splits off and drops steeply into the valley to follow the river on its west side. This road hugs the Kickapoo, running along the bases of high, steep ridges that flank the river and providing more arresting views of the meandering river and the bluffs on its east side. Plum Creek Road goes north a little over three miles before turning sharply west, leaving the valley of the Kickapoo River and ending this tour.

THE LOWER WISCONSIN RIVER

The Wisconsin River, as noted in Chapter 2, is really two different rivers, distinguished by stark differences in their directions of flow and the character of their valleys. The Upper Wisconsin (from the river's source at Lac Vieux Desert to the city of Portage) flows south across hummocky, then rolling, then flat land shaped by glaciers. The Lower Wisconsin River flows west across the state's unique Driftless Area, an ancient craggy landscape never touched by glacial ice. These divergent paths offer clues to what is the most sharply distinguishing difference between the two river segments—their wildly dissimilar geological stories.

The Lower Wisconsin River flows through layers of Cambrian sandstone, the oldest being the Wonewoc Formation, which lies along the river's banks all the way to its mouth on the Mississippi. A slightly younger type of sandstone

5.9 The Lower Wisconsin River Valley viewed from Cactus Bluff in Ferry Bluff State Natural Area

lying over the Wonewoc is the Tunnel City Formation (both of these rock types named for the Wisconsin towns where they were identified). It is nicknamed greensand, because the presence of glauconite, a mineral that formed around the fecal wastes of ancient sea creatures, gives it a green tint. Other layers of younger sandstone are visible on higher parts of the bluffs, and many of the bluffs are capped by dolomite, deposited under early Ordovician seas. These different types of stone vary in their hardness—a factor that played a large role in the formation of the majestic high bluffs that line much of the Lower Wisconsin.

One major difference between the Upper and Lower Wisconsin River Valleys is that the lower valley—formed during the middle of the Quaternary Ice Age—is at least 760,000 years older than the upper valley, which was formed within the last 10,000 years. The latter account is well established by clear evidence left by the glaciers. Less well known is what happened in the older, lower valley, as the clues have become harder to see in the deepening mists of time. Various hypotheses about how the valley formed included notions that have been rejected by most geologists. For example, one story has the ancient Lower Wisconsin River flowing through the Baraboo Hills and carving the notch that now holds Devil's Lake. Another has the river flowing through the Yahara River Valley where Madison sits today.[30]

While these notions have since been discarded, one group of researchers has found intriguing evidence indicating not only that today's Lower Wisconsin River was once separate from the Upper but also that the Lower flowed in the opposite direction—east across the entire state to empty into the ancient St. Lawrence River system.

In 2010, a team of geologists led by Eric C. Carson of the Wisconsin Geological and Natural History Survey studied several features of the Lower Wisconsin that did not fit the then-accepted story of its formation. First, the river narrows dramatically as it flows from Sauk City to its mouth near Prairie du Chien (Figure 5.10). This is unusual for big rivers, which tend to widen as they flow and pick up more tributaries. The river's tributaries offer a second clue. Most of the tributaries in the lower valley hook toward the east as they flow into the

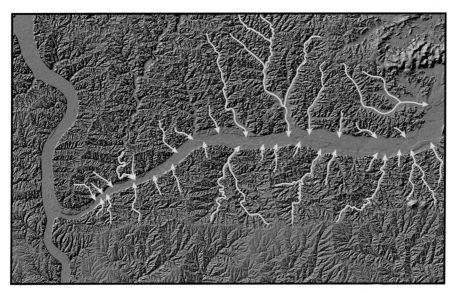

5.10 The Lower Wisconsin River narrows downstream, and its tributaries join the river at unusual angles. COURTESY OF THE WISCONSIN GEOLOGICAL AND NATURAL HISTORY SURVEY (UNIVERSITY OF WISCONSIN–MADISON)

floodplain of the Wisconsin—again, unusual, because tributaries tend to bend toward the direction of flow of the larger river as they approach it. In this case, they should have angled to the west.

The most convincing evidence regarding the river's formation has been gleaned from extensive core drilling conducted by Carson's team, as well as from high-resolution satellite images. The researchers studied a broad stretch of old bedrock near the mouth of the Wisconsin River, called the Bridgeport strath, and discovered that it slopes slightly to the east. This bedrock, made of Tunnel City sandstone, once formed the riverbed. But over hundreds of millennia, the river eroded it, cutting the channel deeper and deeper, so that only a segment of this sandstone remains. It lies about 70 feet above the river and is deeply buried under outwash—sand and gravel carried by earlier, higher versions of the river—which now forms a terrace 100 to 160 feet above the river, one half to one mile wide, and six miles long.[31] The outwash on the terrace slopes west because it was shaped by a postglacial west-flowing river. The inescapable conclusion of Carson's team was that the Bridgeport strath must have been carved by a river flowing to the east for many centuries prior to the time of the last glacier.[32]

Here, then, is the currently favored story of that ancient river—the ancestor of the Lower Wisconsin. Carson and his team call it the Wyalusing River, named for the town of Wyalusing and Wyalusing State Park, both located at the confluence of today's Wisconsin and Mississippi Rivers. That river developed during some stretch of millennia prior to the Ice Age. It followed the course of today's Mississippi, probably from its source to the point where the Wisconsin River now joins it. At that point, the river was forced to veer sharply to the east by a high ridge capped by resistant dolomite standing in its way. From there, the river flowed east in a bed that lay at the level of the Bridgeport strath, moving in the opposite direction of today's Lower Wisconsin. Near Portage, it angled northeast, flowing out of Wisconsin toward the ancient St. Lawrence River, which flowed to the Atlantic. The now-buried valley of the Wyalusing River has been traced through well-drillers' records from the Bridgeport strath to the northeast, more than 180 miles, to the shores of Green Bay.[33]

Meanwhile, the pre-Quaternary Upper Wisconsin River likely flowed in much the same route that it takes today. It meandered from Wisconsin's northern border across a rolling, rocky plain and then curved southeast to join the Wyalusing somewhere east of today's Wausau–Stevens Point area.

The dolomite-capped ridge that blocked the Wyalusing, causing it to turn east, played an important role in the development of what would become the Wisconsin River. The ridge was a remnant of an Ordovician sea floor formed around 450 million years ago. Part of it still exists, in fact—it lies on an east-west line between Madison and the Mississippi River and is known as Military Ridge, now the site of a popular trail. It is matched on the Iowa side of the Mississippi by a high ridge that runs west out of the river valley a short distance into Iowa. As hard as the great ridge's dolomite cap was, it was no match for the slow, steady process of trickling water erosion or the much faster erosion caused by flooding. These processes combined to bore out a major gap in this ridge.

The more recent glaciers—those that arrived during the past 800,000 years—determined the fate of the ancient Wyalusing, Wisconsin, and Mississippi Rivers. These glaciers approached from the northeast flowing southwest, and with each advance, glacial ice blocked the Wyalusing, causing it to back up. Water rose among the steep ridges in the valley, and more than once, the river and tributary valleys became a long, wide, multilegged glacial lake with a sharp bend where the river turned east. Geologists call it Glacial Lake Muscoda (Figure 5.11), and its icy waters threatened to slosh over the tops of the lower ridges as its level rose.

For many thousands of years, rainfall and snowmelt had fed small streams

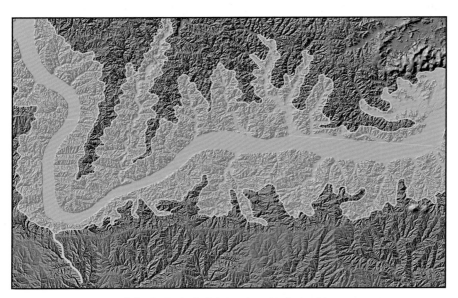

5.11 Glacial Lake Muscoda lay in the bed of the ancient Wyalusing River. COURTESY OF THE WISCONSIN GEOLOGICAL AND NATURAL HISTORY SURVEY (UNIVERSITY OF WISCONSIN–MADISON)

that relentlessly carved little valleys deeper and deeper into the flanks of all the Driftless Area ridges, just as they do today. This erosion was no less steady on the great ridge that blocked the Wyalusing River. One of those steep little tributaries was working away at a point on the south side of the big ridge, slowly dissolving its hard dolomite cap somewhere over the center of today's Mississippi River channel. At the same time, the waters of the rising Glacial Lake Muscoda lapped against the north side of the ridge near its summit. We cannot say with absolute certainty, but it is likely that the lake's waters at some point breached the ridge near where the south-slope tributary's headwaters were eroding the ridge. Here, the rising lake found an exit and began flowing south in the bed of that tributary (Figure 5.11, lower left corner).

From then on, it was simply a matter of time until erosion created a major gap through the high ridge. Once the flowing glacial lake's waters dismantled the dolomite cap, they had no trouble chewing through the softer underlying sandstone layers. A flood of icy glacial lake water gushed through the growing, deepening gap. The undammed river began flowing from the north again and carried centrifugal force as it crashed against the outside of the sharp bend. Water in the eastern leg of Glacial Lake Muscoda (today's Lower Wisconsin Valley) also flowed quickly west through the new drainage point. It was not long before the waters of the Upper Wyalusing had been "pirated," as geologists say, onto a new southward course. With that, the Wyalusing River ceased to be, and the new Mississippi River was born.[34]

What of the lower Wyalusing River, the segment east of the demolished bend in the riverbed? The drainage of Glacial Lake Muscoda must have eroded the east-sloping bedrock to some extent, but possibly not enough to permanently reverse the flow of the lower Wyalusing. When the lake had drained, the river must have been a much smaller stream, flowing more weakly eastward after its main source of water, the upper Wyalusing, was pirated away. It was then supplied only by the tributaries in that lower part of the valley. How and at what point in time did the lower river reverse course and begin to flow west? The answer, as with so many questions about Wisconsin's waterways, is that the glaciers played a major role.

The glaciers that came from the northeast built moraines near the east end of the Lower Wisconsin River Valley. Whenever the climate cooled and the ice advanced a bit farther, it acted as a bulldozer, pushing its accumulated pile of debris forward. When the climate warmed and meltwater steams flowed, they carried much of the debris away from the melting ice front in sheets and torrents, troweling it out in the zone in front of the glacial margin. The effect was that of a road grader, smoothing piles of gravel out and flattening them to make a new roadbed. The glaciers and their meltwaters were easily the largest road graders of all time.

In this way, the eastern part of the lower Wyalusing valley was graded from east to west, the glacial till being thicker nearer the glacial margin and thinning out to the west. Layer upon layer of till was spread across the zone with each advance of the ice sheet. Gradually, a new drainage pattern was established with a continuing, though waning, meltwater flood, along with the flow of the ancient Upper Wisconsin River from the north, pushing water west through the refashioned Wyalusing valley. Ever since the reconfiguration of the rivers, the young Mississippi had been digging its channel deeper, inviting its tributaries to dig deeper to meet it at ever lower levels. Through this down-cutting process, combined with the reshaping of the Wyalusing valley, the young Lower Wisconsin eventually merged at its west end with the young Mississippi River. The final act of the glacial drama in another part of the state helped to firmly establish this new drainage pattern: a catastrophic flood.

When it advanced into the region 19,000 years ago, the Green Bay Lobe blocked the flow of the Upper Wisconsin, the course of which had been established by earlier glaciers, and that river backed up to form Glacial Lake Wisconsin, which occupied an area as large as the Great Salt Lake for more than 3,000 years. When the glacier retreated far enough for the ice dam to break apart, the vast lake drained in a matter of weeks, possibly months—wickedly fast in geologic time. The freed river hauled an unimaginably huge load of sand, gravel, and boulders around the east end of the Baraboo Hills and down into the valley of the new Lower Wisconsin River. This was the final load to be carried by the glacial meltwaters and added to the now westward-sloping bed of the river.

Over centuries, the postglacial flooding diminished steadily. The river took the form of a braided stream with multiple, interconnected channels massaging the sand and gravel across the valley, which had been bored wide and deep by earlier postglacial floods. The sand and gravel filled the valley to depths of more than 150 feet. Whenever the river flooded the valley, it sloshed against the banks, undercutting the sandstone and dolomite masses and causing pieces of these rock layers to plunge from the bluffs into the river. In this way, the steep bluffs that now line the valley were sculpted by the flowing waters below.

As the last postglacial flood diminished, the channels of the broad Lower Wisconsin merged to form a narrower river that began etching a new main channel into the wider riverbed. The more-focused force of the narrower river's flow steadily cut its new trench deeper, eventually leaving broad terraces on either side of the river. With the creation of bluffs, terraces, and a new channel, the postglacial floodwaters put the finishing touches on the Lower Wisconsin River Valley.

When it was stabilized, the Lower Wisconsin River flowed just as it does today, within a broad, steep-sided, flat-bottomed trench, floored with a deep bed of glacial outwash gravel and sand (Figure 5.9). The walls of the trench are 300 to 400 feet high. The river descends just 125 feet from Prairie du Sac to Prairie du Chien—an average of 1.6 feet per mile—and it meanders back and forth irregularly between the valley walls. The channel has split and rejoined many times to form semipermanent wooded islands, and backwater sloughs have formed on the sides of the channel. The river continues to shape the valley

5.12 The Lower Wisconsin River forms many islands and sandbars.

floor, building sandbars and shifting and molding them like a sculptor working with clay (Figure 5.12).[35]

People have been attracted to the Wisconsin River Valley for millennia, finding its terraces good for settlement and the floodplain and bluff forests rich in food resources. Evidence found near the river's mouth south of Prairie du Chien indicates habitation by Paleo-tradition hunters as long as 10,000 years ago. By 8,500 years ago, during the early Archaic period, small family groups of hunter-gatherers were likely migrating through this site to and from the Mississippi Valley. Such camp sites probably existed all along the Lower Wisconsin from Sauk City to the river's mouth. By 3,000 years ago, people had built more permanent settlements in the valley.[36]

Around 2,000 years ago, the Hopewell culture arrived in the area of the Wisconsin River's confluence with the Mississippi, adding the Lower Wisconsin to their extensive trade network. This strengthened an older network that had connected the Fox River to the Lower Wisconsin, thus connecting the St. Lawrence River and the Atlantic Ocean to the Mississippi River and the Gulf of Mexico. Using the rivers as highways, Indigenous traders brought goods to the valley from far and wide, including shells from the Gulf of Mexico, mica from Georgia, obsidian from Wyoming, flint from North Dakota, and copper from the Lake Superior shore.[37] Within several hundred years, the Hopewell were replaced by the Early and Middle Woodland cultures, who added burial mound building to the mix of cultural practices along the Lower Wisconsin.[38]

Around 600 CE, a particular group of Late Woodland people were thriving in the valley. They are noted especially for their effigy mounds, built in the forms of birds, mammals, and reptiles near their villages on the river bluffs and terraces. They continued this practice for at least 700 years, and during this time, they also took up bow-and-arrow hunting and gardening, growing corn, beans, and squash. When European explorers arrived in the lower valley, the dominant groups of the area were the Sac (also called Sauk), Meskwaki (Fox), and Ho-Chunk. Two of the oldest human settlements in the lower valley were at the east end of the valley, where Prairie du Sac is today, and near the mouth of the river on the valley's west end close to Prairie du Chien. According to August Derleth, renowned chronicler of Wisconsin history, the latter settlement was named for the Meskwaki chief Alim, whose name the French interpreted as *chien* (French for *dog*).[39]

Among the early European explorers were Father Jacques Marquette and trader Louis Joliet, who took to the Wisconsin River in 1673. Marquette was credited with first recording the river's name as *Meskousing*, an Algonquian word meaning "stream that meanders through something red," likely referring to the reddish sandstone in the canyon of the Wisconsin Dells in the upper valley. The literature tracks this name as having morphed into *Ouisconsin* by 1718 and *Wisconsin* by the early 19th century. (Note that the state is named after the river, not the other way around.)[40] French explorers and fur traders used the trade network established by Native traders. Were it not for the river, traversing the rugged Driftless Area would have been a major challenge. As a natural highway, the river provided a relatively smooth ride across the area.

In 1766, the British explorer Jonathan Carver traveled the lower valley. At its east end, near present-day Prairie du Sac, he noted a large village occupied by a Sac community, which included medicine man Pyesa, the father of Black Hawk. On the other end of the lower valley, Carver found 300 Meskwaki families living in permanent lodges on the flat river terrace that became known as La Prairies les Chiens. He wrote: "This town is a great mart, where all the adjacent tribes, and even those who inhabit the most remote branches of the Mississippi, annually assemble about the latter end of May, bringing with them their furs to dispose to the traders."[41] Today, the city of Prairie du Chien honors this tradition with its annual Prairie Villa Rendezvous in mid-June—a festival in a recreated frontier village on the old river terrace where Native and French traders celebrated together every year for many decades.

Today, the areas north and south of the Lower Wisconsin River Valley are dominated by agriculture, but the valley itself is well preserved, and much of it probably resembles the valley as it existed before European settlement. This preservation was accomplished partly by the establishment of the Lower Wisconsin State Riverway, a set of valley segments owned by the state, lying between Sauk City and Prairie du Chien. No dams exist below the Prairie du Sac dam, making the entire Lower Wisconsin River the longest free-flowing stretch of river in the Midwest.[42] It provides nearly pristine habitat for turtles, otters, beavers, and numerous fish species, as well as sandhill cranes, eagles, and hawks. The riverway, which enjoys state park status, also protects archaeological sites that help tell the stories of both Indigenous and European settlement in the valley. The Friends of the Lower Wisconsin Riverway do much to support preservation efforts.

Six miles south of Portage on US Highway 51, turn right onto County Road V, which runs west and south to Lake Wisconsin. This widening of the river is caused by the power dam downstream at Prairie du Sac. The lake can be accessed at multiple boat landings and parks. Continue west on Highway V past where it meets State Highway 113 in Okee and then a short distance to County Road VA, which is a spur that takes you to Gibraltar Rock State Natural Area. This 300-foot promontory of dolomite-capped sandstone makes an interesting side trip. It is accessible via a segment of the Ice Age Trail, overlooking the broad flatland and river valley to the south and southeast, respectively. It can also be reached via a hiking trail that traverses high rolling hills, beginning at the ferry landing on the Wisconsin River across from Merrimac.

Backtracking on County Road V and taking Highway 113 west will lead you to the Merrimac Ferry, which shuttles traffic to and from Merrimac as it has for more than a century. From the ferry landing, Highway 188 continues southwest another 10 miles to Prairie du Sac, roughly paralleling the river. Alternatively, you may choose to take the ferry across, pick up State Highway 78 in Merrimac, and follow it west and south to Prairie du Sac. That highway runs along the river through Prairie du Sac and the adjoining town of Sauk City (the pair sometimes referred to as Sauk Prairie). The towns sit on broad outwash plains made of gravel and sand washed out of the Upper Wisconsin Valley by the glaciers' meltwaters. Directly across the river from Sauk Prairie lies the Johnstown Moraine—a high ridge overlooking the river and forming a backdrop for the drama of eagles perching, soaring, and diving as they fish the river during certain times of the year.

From Sauk City, go west on State Highway 60, which runs along the west and north sides of the Wisconsin River almost all the way to its mouth. Also known as the Lower Wisconsin River Road, it was the first to be designated a Wisconsin Scenic Byway—one of five byways chosen for their unique scenic or historical attributes. The first highly recommended side trip off this road is the Ferry Bluff State Natural Area. The one-mile-long Ferry Bluff Road leaves Highway 60 west of Sauk City and ends at a parking area. A trail takes visitors to the top of Cactus Bluff (Figure 5.9), where a set of signs contains information about the geology, biology, and fragility of the bluff ecosystem. From there, hikers can take an unmarked trail northeast to the undeveloped top of Ferry Bluff. Ferry and Cactus Bluffs sit 300 feet above the river and feature remnants of dry prairie. The bluffs provide an important winter roosting area for bald eagles and other raptors, occasionally including the rare peregrine falcon, and they are closed to hikers between November 15 and March 31.

From the Ferry Bluff Road junction, Highway 60 winds its way southwest among the bluffs for another five miles before descending toward the river to thread a course between the bluffs and the river's terraces and floodplain. About 14 miles down this highway is Spring Green, famous for its rich cultural heritage. Continuing west, the road passes several river access points, especially in the small towns of Lone Rock, Gotham, Wauzeka, and Bridgeport near the confluence of the Wisconsin and Mississippi. (Other notable towns on the south side of the river are covered later in this travel guide.) Each of these towns and several other stops along the way have something unique to offer. For example, at the intersection of State Highways 60 and 193, about 17 miles west of Gotham, are the Shadewald Mounds, a set of effigy mounds considered to be among the best preserved in the state and thought to have been built around 1,000 years ago.[43]

Highway 60 ends in Bridgeport where it runs into State Highway 35, which crosses the river just upstream of its confluence with the Mississippi River. For spectacular views of this confluence, cross the river on 35 and go to Wyalusing State Park, where overlooks provide unspoiled views of both the Wisconsin and Mississippi Rivers and of their confluence. In the Green Cloud Picnic Area, overlooking the Mississippi, a large sign provides exceptionally good information on the sites of Native American villages, mounds, and trails all along the Lower Wisconsin River. Here is where the great Military Ridge stretched across the Mississippi Valley hundreds of thousands of years ago before it was picked apart by erosion during the pirating of the Wyalusing River. The park's interior trails provide views of classic Driftless Area topography, leading along the ridge-tops and down into the steep-sided valleys, passing intriguing sandstone caves (Figure 5.13) and outcroppings sculpted by wind and water over millennia.[44]

For close views of the river from its south side, return to County Road C just south of Bridgeport and follow it east. About 14 miles up the river, this road ends on State Highway 133, which continues north and east along the river through the towns of Boscobel, Blue River, Muscoda, and Avoca. Just before 133 crosses the river into Lone Rock, turn right onto State Highway 130 for about two miles

5.13 This cave in Wyalusing State Park, colorfully stained by various minerals in the sandstone, was carved by flowing water.

and turn left onto Iowa County Road C, heading east along the river. It is on this road that Taliesin, Frank Lloyd Wright's famous studio, is located.

Shortly beyond that site, Highway C crosses State Highway 23 and leads to Tower Hill State Park, named for the tower and vertical shaft built in the 1830s for the manufacture of lead shot. The tower sits on a sandstone bluff overlooking the river. A trail there gives hikers splendid views of the valley, as well as a closeup look at the sandstone layers, including the greenish Tunnel City sandstone that was deposited in a quietly deepening inland sea some 500 million years ago.[45] County Road C ends at US Highway 14, and here ends this tour of the Lower Wisconsin River.

THE MISSISSIPPI RIVER

The Mississippi River has earned many monikers in story and song, including the Mighty Mississippi, that Old Man River, and the Father of Waters. It is the

5.14 The Mississippi River Valley viewed from Perrot State Park

largest river, by volume, in North America and the seventh largest in the world. It drains more than 1.2 million square miles of land in 32 states and two Canadian Provinces, its vast watershed sprawling from the Rocky Mountains to the Appalachians.[46] In its current configuration, it is young by geologic standards, at something over 780,000 years of age, but it certainly qualifies as old in human terms.[47]

That third phrase, "Father of Waters," credited to the Ojibwe, is perhaps the most telling. Like a father gathering his children, the great river collects water from countless streams and rivers, draining 40 percent of the continental United States' landmass, including well over half of Wisconsin's land area.[48] It carries these waters all the way to the Gulf of Mexico, where they eventually evaporate and are rained back down into the world's watersheds—a key part of the unending, life-sustaining global water cycle. It seems fitting to tell this river's story in the final chapter of this book on Wisconsin waters.

The Mississippi flows 2,350 miles from its source at Lake Itasca in northern Minnesota to its mouth on Louisiana's Gulf Coast. Wisconsin's share of that length is about 200 miles along the state's western boundary between Prescott, Wisconsin, and Dubuque, Iowa. Over that distance, the river drops a total of 85 feet, an average of five inches per mile, and varies in width from one mile at Prescott to six miles at Trempealeau.[49] Wisconsin's 200 miles of the river's shoreline are unique, because that is the only stretch of the river that lies almost entirely within a broad, steep-sided trench, walled by craggy sandstone bluffs ranging from 200 to 650 feet high. Parts of the river's shorelines in Minnesota and Iowa form the western side of the trench, but nearly all of Wisconsin's share of the river is lined with these beautiful, ancient bluffs (Figure 5.14).

The story of how the Mississippi River Valley was formed begins in Cambrian time with the deposition of deep layers of sandstone interspersed with shale, all overlain by Ordovician dolomite. While the latter was scoured away by erosion over most of Wisconsin, dolomite is still present in the Mississippi River Valley because the western half of Wisconsin slopes to the west and southwest, off the Wisconsin Dome and Arch. Thus, dolomite caps many of the bluffs along the Mississippi and its larger tributaries. The state's bigger rivers have cut trenches through the dolomite, shale, and sandstone bedrock and, especially along the Mississippi, this has made visible a cross section of the ancient layers deposited over tens of millions of years. The Mississippi Valley, therefore, is a window into that distant past.

The Mississippi did not always follow exactly the same course it travels today. The modern Upper Mississippi River was carved by glacial meltwaters beginning about 2.4 million years ago, but before the first glacier came, the environment was quite different. Geologists think the river might have flowed from somewhere in north central Iowa southeast to its present course on the Iowa–Illinois border and then to points south. The earliest glacier covered that ancient valley and pushed its way to what is now Iowa's eastern border, where its meltwaters eventually formed much of the valley south of the Wisconsin–Illinois border.[50]

Meanwhile, north of that border, the same glacier pushed east roughly to the Wisconsin–Minnesota line, where it dammed streams flowing west out of Wisconsin, which then formed icy lakes lapping against the ice wall.[51] When that early glacier began to retreat, its meltwaters flowing along its eastern edge carried the lake waters south, eroding the channel for today's Mississippi. However, instead of flowing down between Wisconsin and Iowa as it does now, that meltwater stream ran into a wall of hard rock and veered east across Wisconsin, creating what is now the Lower Wisconsin River Valley. This is one conclusion drawn by a team of geologists led by Eric C. Carson of the Wisconsin Geological and Natural History Survey who call this ancient stream the Wyalusing River.[52]

Through a long and complicated process, glacial-era floodwaters and frost action eventually eroded a notch in the wall that blocked the Wyalusing River's southern flow (a continuation of the high bulwark now referred to as Military Ridge). The notch was rapidly eroded to form the gap near what is now Wyalusing State Park, and through this gap, the modern Mississippi now flows. Carson's team used extensive research to reach the conclusion that the Wyalusing River existed and was, as geologists say, pirated, or diverted onto the modern course of the Mississippi. One important piece of evidence is that the Mississippi narrows distinctly where the gap was likely carved by the pirated river. Likewise, the tributaries feeding this stretch of the river are short, steep, and narrow, in keeping with the piracy theory.[53] (For more details on this theory, see "The Lower Wisconsin River" earlier in this chapter.)

All of that happened sometime before 780,000 years ago, according to Carson and his fellow researchers. Another more recent course adjustment for the Mississippi—one that took place just 50,000 years ago—created one of Wisconsin's most beautiful settings in Perrot State Park, northwest of La Crosse. The park encompasses a set of unglaciated peaks and ridges isolated from the rest of the Driftless Area by a broad, flat, crescent-shaped area to the north and east of the park (Figure 5.15). At one time, the Mississippi River flowed through this area, and those peaks and ridges were part of the Minnesota bluffs that now rise across the river from the park (Figure 5.14). Over thousands of years, tributaries to the ancient Mississippi carved channels between those peaks and the rest of the Minnesota bluffs.

When one of the earlier glaciers retreated, the Mississippi was carrying loads of sand, silt, and gravel from the melting ice. At the north end of the crescent-shaped area, the Trempealeau River flowed out of a highland into the Mississippi, as it does today. The voluminous flow of glacial debris in the Mississippi, along with sediments from the Trempealeau River, gradually built a low dam across the crescent area, causing the Mississippi to veer southward into those channels carved by tributaries and onto the river's present course.

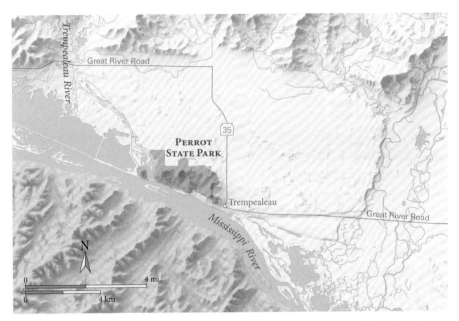

5.15 Perrot State Park is bounded by the Mississippi River on its southwest side and on all other sides by an ancient, abandoned segment of the riverbed. MAPPING SPECIALISTS, LTD., FITCHBURG, WI

This left the Perrot peaks stranded, overlooking the flat area to their north and east and separated from their original home on the Minnesota side by the new course of the Mississippi River.[54]

Whenever a glacier retreated from the Mississippi Valley, water draining off it would course down through the valley carrying untold tons of sediments freed from the melting ice. The meltwater river would take the form of a braided stream—a broad stretch of interwoven channels—troweling glacial sediments across the valley floor in layer upon layer for centuries at a time and raising the level of the floodplain.[55] The renowned Wisconsin geographer Lawrence Martin found an interesting way to describe the result, noting that the 200- to 300-foot-deep accumulation of sediments had "hoisted the river up."[56]

During the last stages of glaciation around 10,000 years ago, great glacial lakes formed against the retreating ice barrier. For the Mississippi River, the most significant of these were Glacial Lake Oshkosh, Glacial Lake Wisconsin, and Glacial Lake Duluth, and Glacial Lake Agassiz, all of which drained to the Mississippi during parts of their existence. The latter was an enormous body of water splayed across all of northwestern Minnesota, eastern North Dakota, and large areas of Ontario, Manitoba, and Saskatchewan. In these and smaller glacial lakes, sediments flowing off the retreating ice would settle out on the lake bottoms, and the water flowing into the river valley would be much clearer than before. Instead of depositing more sediments, these flowing waters tended to cut down through the sediment, forming new and deeper river channels—and all the more so whenever larger volumes of water drained from the lakes.

As the glacial retreat continued, the great glacial lakes would occasionally breach the ice dams that had held them back for centuries at a time and crash down through the stream valleys they had been feeding. Glacial Lake Duluth flooded the St. Croix River; Glacial Lake Wisconsin, the Wisconsin River; and Glacial Lake Agassiz, the ancestral Minnesota River. All of these floods made their way to the Mississippi. (Glacial Lake Oshkosh fed the Mississippi for a time but ultimately drained to Lake Michigan.) Each of these floods widened the valley and cut the main channel of the river deeper.

The Lake Agassiz flood had the biggest effect on the Mississippi during one or more periods between 11,800 and 9,200 years ago.[57] When this or another large glacial impoundment drained, each probably more than once, a huge torrent of ice water overwhelmed the shallow braided channels and bored through the Mississippi Valley. Such floods crashed against the sides of the bluffs, which were made of easily eroded sandstone lying under the more resistant layer of dolomite on the bluff tops. When the water cut deeply enough into a bluff side, the denser, heavier overlying layers would break away from the bluff and plunge into the river. In this way, the dramatic vertical bluffs we see now in the Upper Mississippi Valley were formed (Figure 5.16).

As the glacial lake floods dwindled and the flow of water slowed within braided channels, sediments from the flood once again settled to build up the floor of the river over some number of centuries. Eventually, the waters cleared again and the postglacial Mississippi began to carve its modern channel. The more focused and sediment-free flow cut down into the sediments of the riverbed, dropping into a new trough and leaving broad flat expanses of sand and gravel lying on one or both sides of the channel. These are the terraces on which roads, railroads, and towns have been built. Over the whole history of the river, this flooding-sedimentation-downcutting cycle has occurred numerous times. Today, more than one terrace can be identified along some segments of the river, one slightly higher or lower than the next, like a set of steps, or *benches*—another term for terraces. Today, the terraces along the Mississippi range in height from 40 to 100 feet above the river level.[58]

Another result of the glaciers was the formation of deltas by the larger tributaries to the Mississippi. In some cases, these rivers rolled down to the confluences faster than the Mississippi was flowing, causing the sediments carried by the tributaries to accumulate where they flowed into the Mississippi. These accumulations became partial natural dams that would cause the big river to back up. Some of these streams flowed out of the Driftless Area, washing huge quantities of sand from its deeply eroded valleys. The most prominent example is at the confluence of the Chippewa and Mississippi Rivers. Over time, the Chippewa built a sandy delta that partially blocked the Mississippi, causing the

5.16 Bluffs like these line the Great River Road in Wisconsin. GAIL MARTINELLI

formation of the river's largest pool, today's Lake Pepin—22 miles long by up to 2.5 miles wide. At one time, this lake was more than twice as long, but the Mississippi sediments are gradually filling it in from the upstream end to create broad, river-bottom wetlands.[59]

Other rivers that have built deltas into the Mississippi include the Rush, Trempealeau, and Black Rivers, and each delta is now designated as a state natural area or wildlife refuge. As noted above, the Trempealeau River's delta was partly the cause of a dramatic change of course for the ancient Mississippi River where Perrot State Park is located.

Not long after the last glacier retreated, the Mississippi became a haven for hundreds of plant and animal species assembled within rich river ecosystems. The floodplain first hosted marsh grasses and eventually forests of maple, elm, birch, basswood, cottonwood, and willow. Atop some steep south- and west-facing river bluffs, dry grassland communities took hold. They are called goat prairies because of their steepness. Only goats could safely graze there. Oak savannas grew in the less steep, more shaded areas of the bluff country, as did oak and mixed hardwood forests.

These natural communities eventually hosted dozens of bird species, including great blue herons and egrets that lived in rookeries on the floodplains and hunted frogs and fishes on the edges of river marshes. The Mississippi Valley was and still is used by millions of birds for migration in spring and fall and has been designated one of five major flyways in the United States. For centuries, turkey vultures have soared over the bluffs searching out prey, and bald eagles have fished the river, their prey having been nurtured in the countless backwaters, sloughs, and channels. Also feeding on the fish are river otters that share the bottomland forests with beavers, raccoons, rabbits, foxes, and white-tailed deer. Snakes, frogs, and salamanders also are abundant within these river bottom communities.

Humans have made use of the valley's rich resources since close to 12,000 years ago when Paleo hunters are thought to have arrived in the Upper Mississippi Valley, moving north with the margin of the retreating glacier in search of mammoths, bison, and other big game. According to evidence that dates to around 6,000 years ago, Middle Archaic peoples cultivated sunflowers, corn, squash, beans, and various wild plant species. The Mississippi Valley is one of the earliest known locations for such domestication.[60] Until around 3,000 years ago, people built seasonal camps on the river's terraces for hunting and gathering their food. In the Woodland period, beginning 3,000 years ago, people began to establish more permanent villages, leaving evidence of shelter construction, pottery, weaving, and fishing implements.

The Mississippi served as a major artery within Native American trade networks, beginning around 200 CE. Traders canoeing the river and its tributaries carried not only goods—pottery, pipes, ornaments, and many other items—but also cultural practices that were shared among tribes. The network eventually spanned the area between the Gulf of Mexico and the Great Lakes and from present-day Wyoming to the Appalachians. By around 800 CE, this network included an advanced agrarian society called the Mississippian culture, with stratified social groups, chiefdoms, and towns and cities. The city most well-known to researchers today was Cahokia, located near present-day St. Louis; it was larger in population than London, England, during its peak between 1050 and 1200.[61] The society that lived at the site of today's Aztalan State Park in southeastern Wisconsin was probably built by those Mississippians of Cahokia, looking to expand their culture.

Beginning in the 1600s, European explorers plied the Mississippi, and fur traders soon followed, making use of the trading network already established by Indigenous people. Explorers and traders derived the river's name from the Ojibwe words *Misi ziibi*, interpreted variously as "big river," "long river," and "Father of Waters."[62] For a time, these cultures traded relatively peacefully, often intermarrying. They celebrated together in annual gatherings such as the famous rendezvous in Prairie du Chien, the largest of the many settlements that grew up on the river's broad, ancient terraces.

By the 1850s, Indigenous groups in the Mississippi Valley, primarily the Ojibwe and Ho-Chunk in Wisconsin and the Dakota in Minnesota, had been forced off their lands by the US government. The sites of some of their terrace settlements on the Wisconsin side of the valley are now occupied by several towns, including Cassville, Prairie du Chien, La Crosse, Trempealeau, Pepin, and Prescott.

Heavy trading continues today on the Mississippi River, but the nature of the trading, and of the river itself, have been transformed dramatically by human engineering. Long canoes loaded with goods once quietly traversed the ancient river channels among countless wooded islands. Today, the canoes have been replaced by great rafts of barges pushed and pulled by tugboats through a series of reservoirs created by a lock-and-dam system built in the 1930s. The system converted the Mississippi from a free-flowing river to a series of pools lying between Minneapolis–St. Paul and the Gulf of Mexico. Viewed from the air, the Upper Mississippi bordering Wisconsin displays three differing environments between each set of dams, best described by geologists Robert Dott and John Attig in their seminal book *Roadside Geology of Wisconsin*:

> Immediately above each dam is a large . . . reservoir lake. This passes upstream into an extensive marsh area with backwaters and enclosed sloughs choked with aquatic plants. The third environment, extending up to the next higher dam, has a maze of braided channels, sloughs, and small lakes coursing through bottomland forests. Only this last one preserves the natural, free-flowing character of the river.[63]

That third environment is pictured in Figure 5.17. This pattern is easily recognizable on any set of topographical maps of the valley. With a nine-foot-deep channel dredged regularly by the Army Corps of Engineers, the Mississippi

5.17 Mississippi River bottomland viewed from a bluff-top park overlooking the city of Alma

does facilitate barges and other river traffic, but the lock-and-dam system has dramatically altered the river's ecosystems by raising the water levels on several long stretches and drowning much of the natural floodplain. Dredging combined with waves created by boat traffic have pushed sediments into the backwaters, closing off some side channels. Boat traffic and runoff from adjacent broad farm fields have polluted the Mississippi's water. This has led to the decline of many aquatic plant and animal species and of natural flood control and water filtration, which were once provided by the river and its floodplain.

The Upper Mississippi has received some protections that have helped to restore parts of its natural character. Established in 1924, the Upper Mississippi National Fish and Wildlife Refuge lies between the mouth of the Chippewa River (the south end of Lake Pepin) and Rock Island, Illinois. It encompasses 300 river miles and 240,000 acres of floodplain, comprising marshland, wooded islands, backwater sloughs, oxbow lakes, and sandy beaches. The refuge hosts more than 300 bird species, including half the world's canvasback duck populations, a fifth of all tundra swans in the US, large populations of sandhill cranes and white pelicans, 51 species of mammals, and 119 species of fish.[64]

Some 40 percent of North America's ducks, geese, swans, and other waterfowl use the Mississippi Valley for migration in spring and fall, and they are protected within the Mississippi Flyway by the US Fish and Wildlife Service. This level of protection is an attempt to recognize and honor the Mississippi River Valley, not only as critical habitat for those species but also as a vital element of our entire biosphere.

TRAVEL GUIDE
The Mississippi River

The route for this travel guide is mostly along the Great River Road (GRR), an assemblage of roads tracing the Mississippi River for 3,000 miles through 10 states. Established in 1938, it is the oldest of more than 100 national scenic byways. In Wisconsin, it runs 250 miles and passes through 33 towns and villages. Be prepared for arresting views of the river and its bluffs from the many waysides and overlooks in the river towns.

Wisconsin's segment of the GRR begins on State Highway 35 in Prescott, where the St. Croix River runs into the Mississippi. You may choose to start your trip down the Mississippi by visiting Prescott's GRR Visitor and Learning Center, located on a bluff overlooking the confluence of the St. Croix and Mississippi Rivers. In the 13 miles from Prescott to Diamond Bluff, the road winds among the bluffs east of the river, providing stirring views of coulee country, named for the deep valleys chiseled into sandstone by tributaries of the Mississippi.

Beginning in Diamond Bluff, the road parallels the river, and 10 miles down the road at Bay City, it begins its 24-mile stretch overlooking Lake Pepin, formed thousands of years ago by the growth of the Chippewa River delta. The lake was named for the French explorer Jean Pepin, who settled on its shores in the late 1600s. It constitutes the widest naturally occurring stretch of the Mississippi River. Several vistas can be found along the lakeshore in historic little river towns like Maiden Rock and Stockholm. Maiden Rock is named for a bluff situated about five miles down the road from the town (Figure 5.18) where a wayside sits under a bluff that looms 300 feet above the river. Here, legend has it, a young Sioux woman leaped to her death because tribal elders had forbidden her to marry the one she loved.

Maiden Rock bluff is an example of the window to the past afforded by Wisconsin's Mississippi bluffs. It represents at least 20 million years, from its mid- to late-Cambrian sandstone base to its early Ordovician dolomite cap. Some of its many rock layers contain pellets of the greenish mineral glauconite, formed around 450 million years ago on the floor of a deepening, quiet lagoon. Others hold flat pebbles of sandstone in conglomerate, which tell the story of ferocious storm waves tearing the pebbles from a shallow seashore shortly after the sandstone was deposited. Such are the interpretations of geologists who have studied this cross section of geologic time.[65]

Just down river from Maiden Rock bluff is Stockholm, founded by Swedish immigrants and now a center for folk art and music of the valley. The town of Pepin, just under seven miles down the road from Stockholm, celebrates author Laura Ingalls Wilder, who lived near Pepin as a girl. Her experiences there formed the basis for her first book, *Little House in the Big Woods*. Four miles down the road, the route descends slightly to the level of the Chippewa River at its confluence with the Mississippi. For the next three miles, the highway crosses the broad, lush delta, providing views of its several small lakes and interwoven channels trickling through the dense wetland toward the big river.

On the other side of the delta sits the village of Nelson and, eight miles farther down, the city of Alma. Just northeast of Alma on County Road E, atop the bluff behind the city, is Buena Vista City Park, a manicured green space

5.18 Maiden Rock bluff

with breathtaking views of the town and river valley (Figure 5.17). Here is an excellent place to see what the whole valley looked like before the dams were built—with multiple channels and wooded islands spread across the valley floor. Just upstream is a dam with an artificial lake behind it.

For a closeup look at the valley floor and one of its natural terraces, continue another 14 miles on the GRR to Merrick State Park, a popular fishing destination. This is a secluded little park, thickly wooded, with campsites located on one of the river's islands and two miles of hiking trails. Another 12 miles down the road is a wayside with interesting outcroppings of the valley's oldest sandstone. The tilted layers starkly visible in this outcropping are called crossbedding, and they indicate that sand dunes once lay here, molded by Cambrian-era winds, later to be drowned by an advancing sea and converted to the soft stone you see now.[66]

5.19 A view of undisturbed river bottomland in the Trempealeau National Wildlife Refuge

Just down the GRR is the Trempealeau River crossing, and a mile beyond it, West Prairie Road departs to the south and into the Trempealeau National Wildlife Refuge. This lush preserve allows another opportunity to see the river valley floor as it looked before damming (Figure 5.19). Continuing on the GRR will take you to the village of Trempealeau, which has a proud river town heritage that visitors can experience on its waterfront and in its restaurants and inns.

From Trempealeau, you can take a side trip to Perrot State Park, following the signs along Sullivan Road from the west side of town along the river. The peaks and ridges of this park, as noted earlier, were once part of Minnesota's bluff country across the river. The channel here narrows (Figure 5.15) because the river was funneled into a smaller tributary valley some 50,000 years ago when the main channel north of the park was blocked by the Trempealeau River's delta. This park affords splendid views of the river valley and bluffs (Figure 5.14).[67]

The GRR continues on Highway 35 and proceeds through the cities of Onalaska and La Crosse with good views of Lake Onalaska, created by Lock and Dam 7. In La Crosse, the famous Grandad's Bluff, rising 600 feet above the city, is visible from everywhere and makes a good side trip for a spectacular view of the valley. The river bottom here has retained its natural character—a wildly complex mosaic of islands, sloughs, and braided channels easily seen on another side trip on the State Highway 16 bridge, which goes across the river to La Crescent, Minnesota. Highway 16 crosses Highway 35 heading west from downtown La Crosse.

Stoddard is the next town downstream on the GRR, and about four miles beyond it, Spring Coulee Road leads east to Ramrod Coulee Lane, which takes visitors up the bluff to Ramrod Coulee, an abandoned dolomite quarry and wayside. If the road is closed, a steep hike is involved, but it is worth the effort; the views of the valley from this site are superb, and the quarry, now being reclaimed by the forest, is intriguing. Continuing south, the GRR threads the narrow zone between river and bluffs for 50 miles. The towns of Genoa, DeSoto, Ferryville, and Linxville provide overlooks on the river, with the bluffs looming 400 to 600 feet above much of this route.

At the end of this stretch is the city of Prairie du Chien, the oldest European settlement in the Upper Mississippi Valley and second oldest in Wisconsin. As the most important trading center in the upper valley from the 1600s to the 1800s, this town has a colorful history and a rich river town heritage. Here, on a series of broad terraces made of glacial sediments, Native Americans built a village long before Europeans arrived. The GRR/Highway 35 sits on the highest of the terraces underlain by 200 feet of glacial sediments. US Highway 18 runs west from the GRR, crossing through town as it drops toward the river on "steps" made of succeedingly lower, younger river terraces.

On Highway 18, you can cross the river to Iowa Highway 76 for two interesting side trips. Turning right (north) on 76 takes you to Effigy Mounds National Monument near Marquette, Iowa—one of the best sites in the world for studying Native American effigy mounds. Turning left and going south on 76 for two miles takes you to McGregor, Iowa, the site of Pikes Peak State Park. The park affords panoramic views of the confluence of the Wisconsin and Mississippi Rivers flanked by Wyalusing State Park's high bluff—part of the ridge that once extended across the river and blocked the south-flowing Wyalusing, causing it to flow east across Wisconsin. Here, the modern Mississippi was born when the Wyalusing's flow was pirated through a gap in the ridge more than 780,000 years ago.

Southeast of Prairie du Chien, the GRR crosses the Wisconsin River and intersects with County Road C. Turn right and follow C to County Road X for a must-do side trip to Wyalusing State Park, situated on the high bluff overlooking the Wisconsin–Mississippi confluence. From this bluff, geologists note, you can see how the curve of the valley wall opposite the bluff (south of Prairie du Chien, which is also visible from the bluff) is inconsistent with a river confluence and is more typical of a river flowing around a tight bend, as the ancient Wyalusing must have for many centuries. Eric C. Carson and his team took this to be further evidence of the hypothesized Wyalusing River story.[68] After the last glacier, however, just 10,000 years ago, the Wisconsin River was certainly flowing west into the Mississippi. The view from the park's bluffs at that time would have revealed a tremendous volume of meltwater rolling into the bigger river, creating a massive flow that would course south across the continent.

Wyalusing State Park offers a rich variety of experiences, from paddling the river to exploring a dark cave etched into the bluff's dolomite cap. Throughout

5.20 The valley of the Mississippi River as viewed from Nelson Dewey State Park, with the main river channel in the distance, appears much the same as it did before multiple dams were built on the river.

the park are trails taking hikers into Driftless Area terrain and past sandstone caves and overhangs, some of which are stained with a variety of colors by minerals flowing through the sandstone (Figure 5.13). Other trails take visitors to views of both rivers, as well as the confluence, and to trail signs with information about the park's geology, natural history, and early inhabitants.[69]

The GRR continues south on Highway 35 and turns south onto State Highway 133 just north of Patch Grove. On the north side of Cassville, take County Road VV north from 133 a short distance to Nelson Dewey State Park, which provides stunning bluff-top views of the Mississippi Valley (Figure 5.20). The state park preserves the home of Nelson Dewey, Wisconsin's first governor; a collection of Woodland Indian mounds; and Stonefield, a Wisconsin Historic Site housing a recreated 1890s rural village and agricultural museum.

From Cassville, you can complete a tour of the Wisconsin Mississippi shore by continuing on the GRR/Highway 133 south and east among the bluffs to Potosi, popular for its restored Potosi Brewery. On the south side of town, East Lane Road departs from Highway 133 toward the river. Near this junction, South Main Street runs south on a peninsula jutting into the Mississippi. This narrow gravel road lies on the delta built by the Grant and Potosi Rivers and leads to the Potosi Point Recreation Area. East River Lane Road runs southeast along the river for two miles to the Grant River Recreation Area, a broad riverside green space where you can get one more grand view of the Mighty Mississippi.

ACKNOWLEDGMENTS

My two-year journey exploring the waterways of Wisconsin was not as much a work project as it was an adventure in learning, one that will not end for me even though the writing is done. I would not have been able to begin this adventure without the enthusiastic support of Kate Thompson, director of the Wisconsin Historical Society Press, and her outstanding team of publishing professionals, all of whom have my heartfelt gratitude. In particular, I want to thank Liz Wyckoff, whose bright enthusiasm and careful, thoughtful reading have led to improvements in every chapter of this book. It's a pleasure to work with her and with all the other members of the team.

I will always be grateful to my family—my brothers, John and Rick; sister Amy; their spouses, Rose, Nancy, and Mike; my wife, Gail, and my children—for their enthusiasm and support for this project. John, in particular, gave me written feedback that enriched two chapters. My son, Will, and daughter, Katie, added to the pleasure of this work by coming along on a field trip or two—Katie thinking of good questions to ponder and Will providing good company and some excellent photos.

Thanks go to my sister-in-law Nancy who provided the opening photo (Figure 2.26) in the "Lake Country" section of Chapter 2, taken at our family cabin on Big McKenzie Lake. It is a pleasure to include this image, because "the Lake" has served as a central gathering place for six generations of our family, and early on, it sparked my fascination for Wisconsin's gorgeous waterways. I can't give hearty enough thanks to my departed parents, Art and Betty Spoolman, or to my sister and brother-in-law, Amy and Mike Lanphear, for making that magical place available to us for most of our lives. And to Big McKenzie itself—one of my oldest and dearest friends—unending gratitude.

Gail and I traveled widely in the state and were lucky to have generous friends and relatives who put us up and fed us well in six-star accommodations. For that, thanks to Ted May; Leslie Martinelli and Matt Martinelli; Molly, Angie, and Jamie Ferguson; and Joan Hamblin and David Schiffeling. Warmest thanks go also to our many friends—especially Molly Thoma and Rosie Meinholz—who helped with ideas, information, and feedback at an unforgettable book presentation practice party for my previous book *Wisconsin State Parks: Extraordinary Stories of Geology and Natural History* and since then with their continuing support.

The list of others who provided help with research is long and includes noted geologists and UW–Madison professors David Mickelson and Jean Bahr, UW–Madison geology librarian Marie Dvorzak, Ian Orland with the Wisconsin Geological and Natural History Survey, Joanne Berg and her wonderful crew at Mystery to Me Bookstore, and the Friends Groups at Horicon Marsh and Crex Meadows.

Finally, and certainly most dearly, I thank my research partner, traveling companion, and wife, Gail Martinelli, whose generous moral support, energy, and help with ideas, research, photos, and every other aspect of this work literally kept me going from start to finish. I am grateful for every hour of our time exploring together and look forward to many more.

Notes

Chapter 1

1. Brian Greene, "How Did Water Come to Earth?" *Smithsonian* 44, no. 6 (2013), www.smithsonianmag.com/science -nature/how-did-water-come-to-earth -72037248/; and Lydia J. Hallis et al., "Evidence of Primordial Water in Earth's Deep Mantle," *Science* 350, no. 6262 (November 2015): 795–97.

2. G. Tyler Miller and Scott Spoolman, *Living in the Environment*, 20th ed. (Boston: Cengage Learning, 2021), 56.

3. David M. Mickelson, Louis J. Maher Jr., and Susan L. Simpson, *Geology of the Ice Age National Scenic Trail* (Madison: University of Wisconsin Press, 2011), 17–18.

4. Robert H. Dott Jr. and John W. Attig, *Roadside Geology of Wisconsin* (Missoula, MT: Mountain Press Publishing, 2004), 9.

5. "Wisconsin Water Facts," Wisconsin Water Library, University of Wisconsin Water Resources and Sea Grant Institute, 2020, https://waterlibrary.aqua.wisc.edu/ water-facts/.

6. Gene L. LaBerge, *Geology of the Lake Superior Region* (Tucson: Geoscience Press, 2004), 243.

7. LaBerge, *Lake Superior Region*, 244.

8. Mickelson et al., *Ice Age National Scenic Trail*, 72.

9. David J. Mladenoff, Lisa A. Schulte, and Janine Bollinger, "Broad-Scale Change in the Northern Forests: From Past to Present," chap. 5 in *The Vanishing Present: Wisconsin's Changing Lands, Waters, and Wildlife*, eds. Donald M. Waller and Thomas P. Rooney (Chicago: University of Chicago Press, 2008), 65–66.

10. George I. Quimby, *Indian Life in the Upper Great Lakes, 11,000 BC to AD 1800* (Chicago: University of Chicago Press, 1960), 6–7.

11. "Wisconsin's Name: Where It Came From and What It Means," Historical Essay, Wisconsin Historical Society, www.wisconsin history.org/Records/Article/CS3663.

12. "Old Copper Culture," Milwaukee Public Museum, accessed September 20, 2021, www.mpm.edu/research-collections/anthro

pology/online-collections-research/
old-copper-culture.

Chapter 2

1. Robert Thorson, *Beyond Walton: The Geology, Ecology, and Cultural History of Kettle Lakes from Maine to Montana* (New York: Walker & Company, 2009), 8.
2. Robert H. Dott Jr. and John W. Attig, *Roadside Geology of Wisconsin* (Missoula, MT: Mountain Press Publishing, 2004), 78, 102.
3. Dott and Attig, *Roadside Geology*, 115.
4. Gwen M. Schultz, *Wisconsin's Foundations* (Madison: University of Wisconsin Press, 2004), 165.
5. Dott and Attig, *Roadside Geology*, 59.
6. Lawrence Martin, *The Physical Geography of Wisconsin* (Madison: University of Wisconsin Press, 1965), 474–75.
7. Emmet J. Judziewicz, "Plant Communities of Great Lakes Islands," chap. 9 in *The Vanishing Present: Wisconsin's Changing Lands, Waters, and Wildlife*, eds. Donald M. Waller and Thomas P. Rooney (Chicago: University of Chicago Press, 2008), 116.
8. Margaret Beattie Bogue, *Around the Shores of Lake Superior: A Guide to Historic Sites*, 2nd ed. (Madison: University of Wisconsin Press, 2004), 17.
9. Space limitations prevent inclusion of photos for this and many other Lake Superior area waterfalls, but for excellent descriptions and photos of these falls, see David Hedquist's *Waterfalling in Wisconsin: The Complete Guide to Waterfalls in the Badger State* (Boulder: Trails Books, 2014).

10. Martin, *Physical Geography*, 440.
11. Statistics on the geography of the river and valley are from G. A. Payne, K. E. Lee, G. R. Montz, P. J. Talmage, J. K. Hirsch, and J. D. Larson, *Water-Quality and Aquatic-Community Characteristics of Selected Reaches of the St. Croix River Valley, Minnesota and Wisconsin*, Water-Resources Investigation Report, 02-4147 (US Geological Survey, Washington, DC, 2000), 1.
12. William S. Cordua, *Guidebook for the 10th Annual UW Geologic Field Conference: Geology of the St. Croix Valley, Wisconsin and Minnesota* (River Falls: University of Wisconsin, 1978).
13. Ian S. Williams, ed., "Paleogeography and Structure of the St. Croix River Valley," *Guidebook for the 53rd Annual Tri-State Geological Field Conference* (River Falls: University of Wisconsin, 1989), 5.
14. Mark D. Johnson, Kristin L. Addis, Lisa R. Ferber, Christopher B. Hemstad, Gary N. Meyer, and Laura T. Komai, "Glacial Lake Lind, Wisconsin and Minnesota," *Geological Society of America Bulletin* 111, no. 9 1999): 1371–86.
15. Johnson et al., "Glacial Lake Lind," 1371.
16. Dott and Attig, *Roadside Geology*, 59.
17. National Park Service, St. Croix National Scenic Riverway, updated October 12, 2020, www.nps.gov/sacn/learn/history culture/stories.htm.
18. Alonzo W. Pond, *Interstate Park and Dalles of the St. Croix* (St. Croix Falls, WI: Standard-Press, 1937), 21.
19. Alice Outwater, *Water: A Natural History*

(New York: Basic Books, 1996), 6–17.

20. Thomas F. Waters, *The Streams and Rivers of Minnesota* (Minneapolis: University of Minnesota Press, 1977), 142.

21. For more details on these trails, see Scott Spoolman, *Wisconsin State Parks: Extraordinary Stories of Geology and Natural History* (Madison: Wisconsin Historical Society Press, 2018), 55–63.

22. For more details, see Scott Spoolman, *Wisconsin Rocks! A Guide to Geologic Sites in the Badger State* (Missoula, MT: Mountain Press Publishing Company, 2019), 32–33.

23. Williams, "Paleogeography and Structure," 38.

24. Williams, "Paleogeography and Structure," 32–37.

25. Description in this and next paragraph are based on information from Patty Loew, *Indian Nations of Wisconsin*, rev. 2nd ed. (Madison: Wisconsin Historical Society Press, 2013), 60–63; and Anton Treuer, *Atlas of Indian Nations* (Washington, DC: National Geographic, 2014), 25–27, 107.

26. Doris Green with Michael H. Knight, *Explore Wisconsin Rivers* (Madison: Trails Books, 2008), 9.

27. Martin, *Physical Geography*, 452–63.

28. E. Schneberger and Arthur D. Hasler, "Brule River Survey: Introduction." *Transactions of the Wisconsin Academy of Sciences, Arts and Letters* 36 (1944): 14.

29. Schneberger and Hasler, "Brule River Survey," 17.

30. Nan Wisherd, *Brule River Country* (Brule, WI: Cable Publishing Company, 2017).

31. Wisherd, *Brule River Country*.

32. G. R. Lowry, ed., *Canoeing the Wild Rivers of Northwest Wisconsin* (Spooner, WI: Northwest Canoe Trails, 1977), 4.

33. L. P. Jerrard and R. Jerrard, *The Brule River of Wisconsin*, 2nd ed. (self-published, 2011), 12. Originally published by Hall & Son Printers, Chicago, 1956.

34. Dunn, *Midwest Border River*, 10–14.

35. "The Battle of the Bois Brule River," Dibaajimowin, March 9, 2019, www.dibaajimowin.com/tawnkiyash/the-battle-of-the-bois-brule-river.

36. "Bois Brule River," Travel Superior, www.superiorchamber.org/bois-brule-river.

37. Green, *Explore Wisconsin Rivers*, 9.

38. David M. Mickelson, Louis J. Maher Jr., and Susan L. Simpson, *Geology of the Ice Age National Scenic Trail* (Madison: University of Wisconsin Press, 2011), 304.

39. Dott and Attig, *Roadside Geology*, 132.

40. H. L. Young and S. M. Hindall, *Water Resources of Wisconsin: Chippewa River Basin*, Hydrologic Investigations, Atlas HA3386 (US Geological Survey, 1972).

41. Mickelson et al., *Ice Age National Scenic Trail*, 42.

42. G. W. Andrews, "Late Quaternary Geologic History of the Lower Chippewa Valley, Wisconsin," *Geological Society of America Bulletin* 76, no. 1 (1965): 113–24.

43. D. J. Faulkner, H. M. Jol, and G. L. Running, "Late Quaternary Incision and Terrace Formation in the Lower Chippewa River Valley, West-Central Wisconsin" (poster presentation at Geological Society

of America Meeting, Minneapolis, MN, October 9–12, 2011).

44. Karen Voss and Sarah Beaster, *State of the Lower Chippewa, PUB-WT-554* (Wisconsin Department of Natural Resources, 2001), 26, https://dnr.wi.gov/water/basin/lower chip/lchippewa.pdf.

45. Richard D. Cornell, *The Chippewa: Biography of a Wisconsin Waterway* (Madison: Wisconsin Historical Society Press, 2017), 13.

46. Ronald N. Satz, "Chippewa Treaty Rights: The Reserved Rights of Wisconsin's Chippewa Indians in Historical Perspective," *Transactions of the Wisconsin Academy of Sciences, Arts and Letters* 79, no. 1 (Madison: Wisconsin Academy of Sciences, Arts, and Letters, 1991), 47.

47. Cornell, *The Chippewa*, 60.

48. Wisconsin Department of Natural Resources (WDNR), "Lower Chippewa River State Natural Area," last updated August 31, 2021, https://dnr.wi.gov/topic/lands/naturalareas/index.asp?sna=342.

49. WDNR, *Lower Chippewa River State Natural Area Feasibility Study and Environmental Impact Statement* (Madison: Wisconsin Department of Natural Resources, 2000), 5, 8, 14, 15.

50. Cornell, *The Chippewa*, 203.

51. Dott and Attig, *Roadside Geology*, 97, 153.

52. Paul E. Myers and Douglas R. Maercklein, "Amphibolites and Granites at Jim Falls" (unpublished manuscript, 1978), 1, 2.

53. Cornell, *The Chippewa*, 183.

54. Richard D. Durbin, *The Wisconsin River: An Odyssey through Time and Space* (Cross Plains, WI: Spring Freshet Press, 1997), viii.

55. John W. Attig, "Geologic History of the Wisconsin River" (presentation at the Leopold Center, Baraboo, WI, May 20, 2016). See blog summary on the Aldo Leopold Foundation website, July 1, 2016, www.aldoleopold.org/post/geologic-history-wisconsin-river.

56. Martin, *Physical Geography*, 354.

57. Gwen M. Schultz, *Wisconsin's Foundations* (Madison: University of Wisconsin Press, 2004), 178.

58. Durbin, *Wisconsin River*, 25.

59. Martin, *Physical Geography*, 418.

60. Durbin, *Wisconsin River*, viii.

61. Martin, *Physical Geography*, 353–54.

62. Robert W. Finley, cartographer, *Original Vegetation Cover of Wisconsin* (Madison: Wisconsin Geological and Natural History Survey, 1976).

63. Gene L. LaBerge, *Geology of the Lake Superior Region* (Tucson, AZ: Geoscience Press, 2004), 293–94.

64. All material on American Indians in this section of the chapter is derived from Durbin, *Wisconsin River*, 1–3.

65. Robert Gard, *The Romance of Wisconsin Place Names*, 2nd ed. (Madison: Wisconsin Historical Society Press, 2015), 359; and Michael McCafferty, "On Wisconsin: The Derivation and Referent of an Old Puzzle in American Placenames," *Onoma* 38 (2003), 39–56. See www.wisconsinhistory.org/Records/Article/CS14514.

66. WDNR, Wisconsin Lakes, https://dnr.wi

.gov/lakes/; and WDNR department data, www.dnr.state.wi.us.

67. For more information on Rib Mountain State Park, see Spoolman, *Wisconsin State Parks*, 198–202.

68. LaBerge, *Lake Superior Region*, 244.

69. "Kettle Lakes: A Land Shaped by Icebergs," Ontario Parks blog, July 19, 2019, www .ontarioparks.com/parksblog/kettle-lakes -land-shaped-icebergs/.

70. David J. Mladenoff, Lisa A. Schulte, and Janine Bollinger, "Broad-Scale Change in the Northern Forests: From Past to Present," chap. 5 in *The Vanishing Present: Wisconsin's Changing Lands, Waters, and Wildlife*, eds. Donald M. Waller and Thomas P. Rooney (Chicago: University of Chicago Press, 2008), 65–66.

71. Mickelson et al., *Ice Age National Scenic Trail*, 46.

72. Robert M. Thorson, *Beyond Walden: The Hidden History of America's Kettle Lakes and Ponds* (New York: Walker & Company, 2009), 34.

73. Thorson, *Beyond Walden*, 35.

74. "Lac Court Oreilles, Wisconsin, USA," Lakelubbers, www.lakelubbers.com/lac -courte-oreilles-484/.

75. Gard, *The Romance of Wisconsin Place Names*, 178; and Loew, *Indian Nations of Wisconsin*, 71.

76. Information from the David R. Obey Ice Age Interpretive Center, Chippewa Moraine State Recreation Area, New Auburn, Wisconsin.

77. Dott and Attig, *Roadside Geology*, 72–73.

78. Carl Fries, Jr., *Geology and Groundwater of the Trout Lake Region, Vilas County, Wisconsin*, Open-File Report 36-1 (Wisconsin Geological and Natural History Survey, 1936); and Dott and Attig, *Roadside Geology*, 73–74.

79. Ian K. Freeman, Rebecca J. Moore, and Kent M. Syverson, "Geology Hiking Guide for the Ice Age Trail, Straight Lake State Park Area, Polk County, Wisconsin" (University of Wisconsin–Eau Claire Department of Geology, 2012). For more details on this park, see Spoolman, *Wisconsin State Parks*, 52–55.

80. Mark D. Johnson, "Pleistocene Geology of Barron County, Wisconsin," Information Circular 55 (Madison: Wisconsin Geology and Natural History, 1986), 28.

Chapter 3

1. Robert H. Dott Jr. and John W. Attig, *Roadside Geology of Wisconsin* (Missoula, MT: Mountain Press Publishing, 2004), 297.

2. John McPhee, *Annals of the Former World* (New York: Farrar, Straus and Giroux, 1981), 31–32.

3. J. K. Greenberg, B. A. Brown, L. G. Medaris, and J. L. Anderson, "The Wolf River Batholith and Baraboo Interval in Central Wisconsin," *Field Trip Guidebook Number 12, Thirty-Second Annual Meeting Institute on Lake Superior Geology* (Madison: Wisconsin Geology and Natural History Survey, 1986), 2.

4. Dott and Attig, *Roadside Geology*, 102.

5. Carl E. Dutton, *Geology of the Florence*

Area, Wisconsin and Michigan, USGS Professional Paper 633 (Washington, DC: US Government Printing Office, 1971), 2.

6. J. A. Clark, K. M. Befus, T. S. Hooyer, P. W. Stewart, T. D. Shipman, C. T. Gregory, D. J. Zylstra, "Numerical Simulation of the Paleohydrology of Glacial Lake Oshkosh, Eastern Wisconsin," abstract, *Quaternary Research* 69, no. 1 (January 2008).

7. Dutton, *Geology of the Florence Area*, 2.

8. Dott and Attig, *Roadside Geology*, 287.

9. "Amidst the Daily Tasks: The Long View," River Alliance of Wisconsin, March 14, 2017, https://wisconsinrivers.org/tag/sturgeon.

10. Patty Loew, *Indian Nations of Wisconsin*, rev. 2nd ed. (Madison: Wisconsin Historical Society Press, 2013), 23.

11. "History of Marinette," Marinette Expanding Horizons, www.marinette.wi.us/261/History-of-Marinette.

12. Loew, *Indian Nations*, 23.

13. J. Waldrip, "Lake Sturgeon Passage at Five Hydroelectric Dams on the Menominee River," *International Conference on Engineering and Ecohydrology for Fish Passage 2014*, 3.

14. US Fish and Wildlife Service, "The Story of Lake Michigan Sturgeon: The Menominee River," Great Lakes Restoration Initiative, 2010, www.fws.gov/GLRI; and Eric Peterson, "Sturgeon Elevator Tour at Menominee Dam," report on WLUK Fox 11 News, May 22, 2018, https://fox11online.com/news/local/sturgeon-elevator-tour-at-menominee-dam.

15. "Marinette, Queen [Marinette Chevalier Jacobs] 1793–1865," Wisconsin Historical Society, www.wisconsinhistory.org/Records/Article/CS1686.

16. C. E. Boyd, "Geology and Geography of Marinette County" (PhD thesis, University of Wisconsin, Madison, 1914), 8–9.

17. Dott and Attig, *Roadside Geology*, 288; and Rachel K. Paull and Richard A. Paull, *Wisconsin and Upper Michigan*, K/H Geology Field Guide Series (Dubuque, IA: Kendall/Hunt Publishing Company, 1980), 161.

18. Boyd, "Geology and Geography," 4.

19. Robert Gard, *The Romance of Wisconsin Place Names*, 2nd ed. (Madison: Wisconsin Historical Society Press, 2015), 322. See also R. E. Gard and L. G. Sorden, *Wisconsin Lore* (New York: Duell, Sloan & Pearce, 1966), 186–87.

20. Jean M. Olson, "The Geology of the Lower Proterozoic McCaslin Formation, Northeastern Wisconsin," *Geoscience Wisconsin* 9 (June 1984): 1.

21. J. W. Attig and W. G. Batten, cartographers, *Quaternary Geology of the Peshtigo River State Forest Area*, Open-File Report 2004-05 (Madison: Wisconsin Geological and Natural History Survey, 2004).

22. Some of the descriptive details about these waterfalls were gleaned from David Hedquist's *Waterfalling in Wisconsin: the Complete Guide to Waterfalls in the Badger State* (Boulder: Trails Books, 2014).

23. Dott and Attig, *Roadside Geology*, 101.

24. W. W. Simpkins, M. C. McCartney, and D. M. Mickelson, *Pleistocene Geology of*

Forest County, Wisconsin, Information Circular 61 (Madison: Wisconsin Geological and Natural History Survey, 1987), 3.

25. Dott and Attig, *Roadside Geology*, 117.

26. D. M. Mickelson, *Glacial and Related Deposits of Langlade County, Wisconsin*, Information Circular 52 (Madison: Wisconsin Geological and Natural History Survey, 1986), 3.

27. Wisconsin Department of Natural Resources (WDNR), "Central Lake Michigan Coastal Ecological Landscape," PUB-SS-1131J 2015, chap. 8 in *The Ecological Landscapes of Wisconsin: An Assessment of Ecological Resources and a Guide to Planning Sustainable Management* (Madison: WDNR, 2015), J-10.

28. P. G. Olcott, cartographer, *Water Resources of Wisconsin, Fox-Wolf Rivers Basin*, Hydrologic Investigations Atlas HA-321 (Madison: Wisconsin Geological and Natural History Survey, 1968).

29. WDNR Watersheds, WDNR GIS Open Data, https://data-wi-dnr.opendata.arcgis.com/datasets/watersheds.

30. Charles H. Velte, *Historic Lake Poygan* (self-published, 1976), 10.

31. Shawano Dam history noted in https://moviecultists.com/where-is-the-shawano-dam, 2021; US Fish and Wildlife Service, Great Lakes Lake Sturgeon Web Site, www.fws.gov/midwest/sturgeon/biology.htm.

32. Clarence J. Milfred, Gerald W. Olson, and Francis D. Hole, *Soil Resources and Forest Ecology of Menominee County, Wisconsin*, Bulletin 85, Soil Series No. 60 (Madison: Wisconsin Geological and Natural History Survey, 1967), 20.

33. Milfred et al., *Soil Resources*, 21.

34. Loew, *Indian Nations*, 28–33.

35. Milfred et al., *Soil Resources*, 22.

36. Menominee Tribal Enterprises, "The Forest Keepers," www.mtewood.com.

37. Loew, *Indian Nations*, 38.

38. Ryan Koenigs and Ronald Bruch, "Lake Sturgeon Reintroduction and Movement in the Upper Wolf River, Wisconsin," abstract (American Fisheries Society, 144th Annual Meeting, August 2014).

39. John A. Luczaj, "Geology of the Niagara Escarpment in Wisconsin," *Geoscience Wisconsin* 22, Part 1 (2013): 6–9, 16.

40. Luczaj, "Niagara Escarpment," 21.

41. Luczaj, 21.

42. Robert R. Schrock, "Geology of Washington Island and its Neighbors, Door County, Wisconsin," *Transactions of the Wisconsin Academy of Science, Arts, and Letters* 32 (1940): 199–227.

43. Luczaj, "Niagara Escarpment," 24.

44. Lawrence Martin, *The Physical Geography of Wisconsin* (Madison: University of Wisconsin Press, 1965), 239.

45. Luczaj, "Niagara Escarpment," 25–26.

46. Martin, *Physical Geography*, 241.

47. "Niagara Escarpment," Door County Coastal Byway, https://doorcountycoastalbyway.org/niagara-escarpment/. Karst erosion presents a serious challenge to today's growing population on the Door Peninsula in controlling pollution and contamination of drinking water.

48. Dott and Attig, *Roadside Geology*, 249.

49. Luczaj, "Niagara Escarpment," 26.

50. Howard Deller and Paul Stoelting, "Wisconsin's Door Peninsula and Its Geomorphology" (1986), 40, wisconsingeography.files.wordpress.com/2013/05/1986-volume-2-wisconsins-door-peninsula-and-its-geomorphology-by-howard-deller.pdf.

51. Patrick Jung, *The Misunderstood Mission of Jean Nicolet* (Madison: Wisconsin Historical Society Press, 2018), 130–132.

52. Anton Treuer, *Atlas of Indian Nations* (Washington, DC: National Geographic, 2014), 51.

53. Martin, *Physical Geography*, 220.

54. "Copper Culture State Park," Oconto Historical Society, www.ocontoctyhistsoc.org/copper-culture-state-park.

55. WDNR, "Red Banks Alvar State Natural Area," https://dnr.wi.gov/topic/Lands/naturalareas/index.asp?SNA=332.

56. Dott and Attig, *Roadside Geology*, 314.

57. "Niagara Escarpment."

58. Luczaj, "Niagara Escarpment," 7.

59. Dott and Attig, *Roadside Geology*, 302.

60. Luczaj, "Niagara Escarpment," 7.

61. Luczaj, 21.

62. Martin, *Physical Geography*, 238.

63. The story of the stages of Lake Michigan was compiled from several sources: Dott and Attig, *Roadside Geology*, 248–50; Gwen M. Schultz, *Wisconsin's Foundations* (Madison: University of Wisconsin Press, 2004), 164–65; Todd A. Thompson, "After the Thaw: The Development of Lake Michigan," Indiana Geological and Water Survey, https://igws.indiana.edu/FossilsAndTime/LakeMichigan; William R. Farrand, *The Glacial Lakes around Michigan*, Bulletin 4 (Lansing: Michigan Department of Environmental Quality, 1988), 9; and Robb Gillespie, William B. Harrison III, and G. Michael Grammer, *Geology of Michigan and the Great Lakes* (San Francisco: Cengage Learning, 2008), 10–12.

64. WDNR, "Central Lake Michigan," J-10.

65. Dott and Attig, *Roadside Geology*, 248–50.

66. Farrand, *Glacial Lakes around Michigan*, 9.

67. Deller and Stoelting, "Wisconsin's Door Peninsula," 35.

68. David Tenenbaum, "As Lake Michigan Rises, Bluffs Collapse and Geologists Explore," *University of Wisconsin-Madison News* (February 15, 2019), https://news.wisc.edu/as-lake-michigan-rises-bluffs-collapse-and-geologists-explore/.

69. Material in this and the three preceding paragraphs is derived from Loew, *Indian Nations*, chapters 4 and 6.

70. Gard, *Romance of Wisconsin Place Names*, 212.

71. Jack W. Travis, "An Overview of the 2009 Geology Field Conference of the Wisconsin Section of the American Institute of Professional Geologists," in *Field Trip Guidebook: Geology of Brown and Door Counties, Wisconsin* (Denver, CO: American Institute of Professional Geologists), 5, www.aipgwisconsin.org/jpegs/Guide%20Book.pdf.

72. Dott and Attig, *Roadside Geology*, 314.

73. WDNR, "Cave Point-Clay Banks State Natural Area," https://dnr.wi.gov/topic/Lands/naturalareas/index.asp?SNA=559.

74. Kewaunee County History blog, https://kewauneecountyhistory.blogspot.com/.

75. Dott and Attig, *Roadside Geology*, 304.

Chapter 4

1. Robert H. Dott Jr. and John W. Attig, *Roadside Geology of Wisconsin* (Missoula, MT: Mountain Press Publishing, 2004), 239.

2. Dott and Attig, *Roadside Geology*, 20–21, 260.

3. Wisconsin Department of Natural Resources (WDNR), "Central Lake Michigan Coastal Ecological Landscape," PUB-SS-1131J 2015, chap. 8 in *The Ecological Landscapes of Wisconsin: An Assessment of Ecological Resources and a Guide to Planning Sustainable Management* (Madison: WDNR, 2015), J-9.

4. G. Tyler Miller and Scott Spoolman, *Living in the Environment*, 20th ed. (Boston: Cengage Learning, 2021), 153.

5. Milwaukee Public Museum (MPM), "The Virtual Silurian Reef," www.mpm.edu/plan-visit/educators/resources-development/virtual-silurian-reef.

6. MPM, "The Virtual Silurian Reef."

7. Gwen M. Schultz, *Wisconsin's Foundations* (Madison: University of Wisconsin Press, 2004), 18.

8. Dott and Attig, *Roadside Geology*, 20–21, 259.

9. Lawrence Martin, *The Physical Geography of Wisconsin* (Madison: University of Wisconsin Press, 1965), 285.

10. Schultz, *Wisconsin's Foundations*, 161.

11. Schultz, *Wisconsin's Foundations*, 161.

12. Dott and Attig, *Roadside Geology*, 261.

13. Robert Gard, *The Romance of Wisconsin Place Names*, 2nd ed. (Madison: Wisconsin Historical Society Press, 2015), 215.

14. Dott and Attig, *Roadside Geology*, 268–69.

15. D. G. Mikulic and J. Kluessendorf, "Wauwatosa's Ancient Reef and Amateur Naturalist: Dedication of the Schoonmaker Reef and the Fisk Holbrook Day Home as National Historic Landmarks: August 21, 1998," abstract (Wauwatosa, WI: Wauwatosa Historical Society, 1998).

16. Dott and Attig, *Roadside Geology*, 267.

17. WDNR, "Chiwaukee Prairie State Natural Area," https://dnr.wi.gov/topic/lands/natural areas/index.asp?SNA=54

18. See "Wolf River" in Chapter 3. The Wolf and Fox Rivers are considered to be of one system because they share a watershed defined by the ancient lake bed.

19. US Geological Survey, The National Map, National Hydrography Dataset, last updated August 2021, https://viewer.nationalmap.gov/viewer.

20. Dott and Attig, *Roadside Geology*, 283–85.

21. Joanne Kluessendorf and Donald G. Mikulic, "The Lake and the Ledge: Geological Links between the Niagara Escarpment and Lake Winnebago," *65th Annual Tri-State Geological Field Conference: 2–3 October, 2004* (Menasha, WI: Weis Earth Science Museum, 2004), 6.

22. Kluessendorf and Mikulic, "Lake and the Ledge," 44–46.

23. Martin, *Physical Geography*, 355.

24. DeLorme Mapping Co., *Wisconsin Atlas & Gazetteer*, 14th ed. (Yarmouth, ME: Garmin Ltd., 2017), 78.

25. Martin, *Physical Geography*, 358.

26. Charles H. Velte, *Historic Lake Poygan* (self-published, 1976), 11.

27. P. G. Olcott, cartographer, *Water Resources of Wisconsin, Fox-Wolf Rivers Basin*, Hydrologic Investigations Atlas HA-321 (Madison: Wisconsin Geological and Natural History Survey, 1968).

28. Martin, *Physical Geography*, 283.

29. WDNR, *Lower Fox River Basin Integrated Management Plan*, Publ. WT-666-2001 (Madison: WDNR, 2001), 27.

30. Mike Hoffman, "Menominee Place Names in Wisconsin," The Menominee Clans Story, https://www4.uwsp.edu/museum/menomineeclans/places.

31. Martin, *Physical Geography*, 355.

32. David M. Mickelson, Louis J. Maher Jr., and Susan L. Simpson, *Geology of the Ice Age National Scenic Trail* (Madison: University of Wisconsin Press, 2011), 228.

33. Mickelson et al., *Ice Age National Scenic Trail*, 230–32.

34. WDNR, "Page Creek Marsh State Natural Area," https://dnr.wi.gov/topic/lands/natural areas/index.asp?SNA=330.

35. WDNR, "White River Marsh Wildlife Area," https://dnr.wi.gov/topic/lands/wild lifeareas/whiteriver.html.

36. Weis Earth Science Museum website, https://uwosh.edu/weis/.

37. Aldo Leopold, *A Sand County Almanac* (Oxford, England: Oxford University Press, 1949), 101.

38. Dott and Attig, *Roadside Geology*, 14–19.

39. Janet R. Battista, "Quaternary Geology of the Horicon Marsh Area" (master's thesis, University of Wisconsin–Madison, 1990), 10.

40. Battista, "Quaternary Geology," 9, 25, 27.

41. Battista, "Quaternary Geology," 34.

42. Battista, "Quaternary Geology," 34.

43. Battista, "Quaternary Geology," i.

44. Battista, "Quaternary Geology," 95.

45. US Fish and Wildlife Service (USFWS), "Horicon National Wildlife Refuge," www.fws.gov/refuge/Horicon/wildlife_and_habitat/index.html.

46. USFWS, "Horicon National Wildlife Refuge."

47. WDNR, "Human History of Horicon Marsh Wildlife Area," https://dnr.wi.gov/topic/lands/WildlifeAreas/horicon/humhist.html.

48. WDNR, "Human History of Horicon Marsh."

49. USFWS, "Horicon Marsh Wildlife Refuge."

50. Leopold, *Sand County Almanac*, 108.

51. Ron Shutvet, "Geologic History of the Olin-Turville Parks Area," Friends of Olin-Turville, www.olin-turville.org/home.

52. Shutvet, "Olin-Turville Parks Area"; and Eric Booth, "Thinking Deep: The Short Story of Yahara's Geologic Past," May 28, 2014, http://yaharawsc.wordpress.com/2014/05/28/thinking-deep-the-short-story-of-yaharas-geologic-past.

53. Lee Clayton and John W. Attig, *Pleistocene Geology of Dane County, Wisconsin*, Bulletin 95 (Madison: Wisconsin Geological and Natural History Survey, 1997), 47, 50.

54. David M. Mickelson and Susan L. Hunt,

Landscapes of Dane County, Wisconsin, Educational Series 43 (Madison: Wisconsin Geological and Natural History Survey, 2007), 22.

55. Robert Thorson, *Beyond Walton: The Geology, Ecology, and Cultural History of Kettle Lakes from Maine to Montana* (New York: Walker & Company, 2009), 34.

56. Clayton and Attig, *Pleistocene Geology*, 60.

57. Mickelson and Hunt, *Landscapes of Dane County*, 22.

58. Booth, "Thinking Deep."

59. Clayton and Attig, *Pleistocene Geology*, 52.

60. Martin, *Physical Geography*, 271–76.

61. Martin, *Physical Geography*, 223.

62. Mickelson and Hunt, *Landscapes of Dane County*, 30–31; Clayton and Attig, *Pleistocene Geology*, 60.

63. Mickelson and Hunt, *Landscapes of Dane County*, 32; Clayton and Attig, *Pleistocene Geology*, 33.

64. Dott and Attig, *Roadside Geology*, 277; Gard, *Romance of Wisconsin Place Names*, 208–10, 218.

65. Livia Gershon, "Intact 1,200-Year-Old Canoe Recovered from Wisconsin Lake," *Smithsonian Magazine* (November 8, 2021), https://www.smithsonianmag.com/ smart-news/1200-year-old-canoe-found -in-wisconsin-lake-180979024/.

66. Mark Gajewski, "Original Names of the Four Lake Region," *Historic Madison*, www .historicmadison.org/originalnames; and Martin, *Physical Geography*, 276.

67. Gard, *Romance of Wisconsin Place Names*, 170.

68. Gajewski, "Original Names."

69. Schultz, *Wisconsin's Foundations*, 86.

70. Mickelson and Hunt, *Landscapes of Dane County*, 24–29.

71. For the detailed history of limnology in Wisconsin, see Scott Spoolman, "Expanding Waters: How Wisconsin Became the Wellspring of a New Scientific Field," *Wisconsin Magazine of History* 86, no. 4 (Summer 2003): 16–29.

72. Kenneth Casper, "Walking Governor's Island in Madison, WI," Wisconsin Explorer, https://wisconsin-explorer.blogspot.com/ 2018/08/walking-governors-island-in -madison-wi.html.

73. City of Madison Parks Division, "Cherokee Marsh Conservation Park–North," www.cityofmadison.com/parks/cherokee north/; "Cherokee Marsh Conservation Park–South," www.cityofmadison.com/ parks/cherokeesouth; and "Cherokee Marsh Conservation Park–Mendota," www.cityofmadison.com/parks/find-a -park/park.cfm?id=1183.

74. Mickelson and Hunt, *Landscapes of Dane County*, 29.

75. Mickelson and Hunt, *Landscapes of Dane County*, 21.

76. Mark Wegener, Paul Zedler, Brad Herrick, and Joy Zedler, "Curtis Prairie: 75-Year-Old Restoration Research Site," Leaflet 16, UW Arboretum Leaflets, August 2008, https://arboretum.wisc.edu/content/up loads/2015/04/16_ArbLeaflet.pdf.

77. Mickelson and Hunt, *Landscapes of Dane County*, 25.

78. WDNR, "Waubesa Wetlands State Natural

Area," https://dnr.wi.gov/topic/Lands/naturalareas/index.asp?SNA=114.

79. County of Dane Wisconsin, "Fish Camp County Park," https://parks-lwrd.countyofdane.com/park/FishCamp.

80. WDNR, "Lake Kegonsa State Park," https://dnr.wi.gov/topic/parks/name/lakekegonsa/nature.html.

81. Mickelson and Hunt, *Landscapes of Dane County*, 25.

Chapter 5

1. Kenneth I. Lange, *Song of Place: A Natural History of the Baraboo Hills* (Baraboo, WI: Ballindalloch Press, 2014), 497.

2. Robert H. Dott Jr. and John W. Attig, *Roadside Geology of Wisconsin* (Missoula, MT: Mountain Press Publishing, 2004), 194.

3. Gene L. LaBerge, *Geology of the Lake Superior Region* (Tucson: Geoscience Press, 2004), 117–18.

4. This and the next five paragraphs contain information from Dott and Attig, *Roadside Geology*, 194, 200, 215–22.

5. Gwen M. Schultz, *Wisconsin's Foundations* (Madison: University of Wisconsin Press, 2004), 160.

6. Lange, *Song of Place*, 88.

7. Lange, *Song of Place*, 10, 158.

8. See Lange, *Song of Place*, 393–96, 456–57, for exhaustive lists and descriptions of species thriving or just surviving in the Baraboo Hills.

9. Dott and Attig, *Roadside Geology*, 216.

10. Lawrence Martin, *The Physical Geography of Wisconsin* (Madison: University of Wisconsin Press, 1965), 195.

11. Curt Meine, "The View from Man Mound," chap. 2 in *The Vanishing Present: Wisconsin's Changing Lands, Waters, and Wildlife*, eds. Donald M. Waller and Thomas P. Rooney (Chicago: University of Chicago Press, 2008), 22; and Gary S. Casper, "Changes in Amphibian and Reptile Communities," chap. 20 in Waller and Rooney, *Vanishing Present*, 287.

12. David Rogers, Thomas P. Rooney, and Rich Henderson, "From the Prairie-Forest Mosaic to the Forest: Dynamics of Southern Wisconsin Woodlands," chap. 7 in Waller and Rooney, *Vanishing Present*, 91.

13. Waller and Rooney, *Vanishing Present*, plate 3.

14. Kickapoo Valley Reserve, "About Us," http://kvr.state.wi.us/About-Us/History/Archaeology/.

15. James B. Stoltman and Robert F. Boszhardt, "An Overview of Driftless Area Prehistory," chap. 6 in *The Physical Geography and Geology of the Driftless Area*, Special Paper 543, eds. Eric C. Carson, J. Elmo Rawling III, J. Michael Daniels, and John W. Attig (Boulder: Geological Society of America, 2019), 97.

16. Driftless Wisconsin, https://driftlesswisconsin.com.

17. Patty Loew, *Indian Nations of Wisconsin*, rev. 2nd ed. (Madison: Wisconsin Historical Society Press, 2013), 45–46. Some popular sources say the river is named for a group of Kickapoo Indians. There are limited reports of a people called the Kickapoo

that lived in the river valley before 1700. The homeland for the Kickapoo people is thought to be along the Wabash River in the area around what is now Terre Haute, Indiana. Like the Kickapoo River, the Kickapoo people are thought to have been named for their wandering, as a nomadic tribe; perhaps those unknown people who named the river were familiar with, and reminded of, the Kickapoo people. Today, as a result of their own movements and those coerced by the US government, their descendants have established communities in Texas, Oklahoma, and Kansas. See also Curt Meine and Keefe Keeley, eds., *The Driftless Reader* (Madison: University of Wisconsin Press, 2017).

18. Waller and Rooney, *Vanishing Present*, plates 3 and 4.

19. Kickapoo Valley Reserve, "Visitor Guide," updated January 2020, http://kvr.state .wi.us/Documents/Maps/Recreation Maps/2020%20Visitor%20Guide.pdf.

20. US Department of Agriculture, National Resource Conservation Service, *Rapid Watershed Assessment, Kickapoo Valley Watershed*, Report No. HUC: 07070006, July 2008.

21. Joy B. Zedler and Kenneth W. Potter, "Southern Wisconsin's Herbaceous Wetlands: Their Recent History and Precarious Future," chap. 15 in Waller and Rooney, *Vanishing Present*, 200.

22. Colin S. Belby, Lindsay J. Spigel, and Faith A. Fitzpatrick, "Historic Changes to Floodplain Systems in the Driftless Area,"

chap. 7 in Carson et al., *Physical Geography and Geology of the Driftless Area*, 125–26.

23. Wisconsin Department of Natural Resources (WDNR), "Kickapoo River Wildlife Area–Bell Center Unit," https://dnr.wi.gov/ topic/lands/wildlifeareas/kickapoobcu.html.

24. Kickapoo Valley Reserve, "Natural Features," http://kvr.state.wi.us/Natural -Features/Land-Cover.

25. Kickapoo Valley Reserve, "Natural Features."

26. Kickapoo Valley Reserve, "Visitor Guide."

27. WDNR, "Mount Pisgah Hemlock-Hardwoods State Natural Area," https://dnr.wi .gov/topic/Lands/naturalareas/index.asp ?SNA=15.

28. WDNR, "Kickapoo River Wildlife Area."

29. WDNR, "Kickapoo River Wildlife Area."

30. Martin, *Physical Geography*, 189–92.

31. Martin, *Physical Geography*, 189.

32. Eric C. Carson and J. Elmo Rawling III, "Late Cenozoic Evolution of the Lower Wisconsin River Valley: Evidence for the Reversal of the River," *Guidebook for the 2015 Geological Society of America North-Central Section Meeting Field Trip #5* (Madison: Wisconsin Geological and Natural History Survey, 2015), 5–9.

33. Carson and Rawling, "Late Cenozoic Evolution," 7.

34. Carson and Rawling, "Late Cenozoic Evolution," 10.

35. Martin, *Physical Geography*, 418, 185–89; and Dott and Attig, *Roadside Geology*, 147.

36. Richard D. Durbin, *The Wisconsin River: An Odyssey through Time and Space* (Cross

Plains, WI: Spring Freshet Press, 1997), 2.

37. WDNR, "Lower Wisconsin State River-way," https://dnr.wisconsin.gov/topic/lands/lowerwisconsin/history.

38. Durbin, *Wisconsin River*, 2.

39. August Derleth, *The Wisconsin: River of a Thousand Isles* (Madison: University of Wisconsin Press, 1942), 28.

40. Robert Gard, *The Romance of Wisconsin Place Names*, 2nd ed. (Madison: Wisconsin Historical Society Press, 2015), 359; and Michael McCafferty, "On Wisconsin: The Derivation and Referent of an Old Puzzle in American Placenames," *Onoma* 38 (2003): 39–56. See Wisconsin Historical Society, www.wisconsinhistory.org/Records/Article/CS14514.

41. Derleth, *The Wisconsin*, 29.

42. WDNR, "Lower Wisconsin State Riverway Maps," https://dnr.wisconsin.gov/topic/lands/lowerwisconsin/maps.

43. The mounds were named for Frank Shade-wald (1932–2013) who purchased the land at this junction with the intent of preserving these ancient earthworks for all to enjoy. Upon his death, the land passed to the Three Eagles Foundation, Inc., which continues to maintain and protect the mounds. See http://3-eagles.org/.

44. For more details on Wyalusing State Park, see Scott Spoolman, *Wisconsin State Parks: Extraordinary Stories of Geology and Natural History* (Madison: Wisconsin Historical Society Press, 2018), 99–105.

45. Spoolman, *Wisconsin State Parks*, 83–87.

46. National Park Service, "Mississippi River Facts," updated September 28, 2021, www.nps.gov/miss/riverfacts.htm.

47. Carson and Rawling, "Late Cenozoic Evolution," 11.

48. WDNR, "Gateway to Wisconsin's Basins and Watersheds," https://dnr.wisconsin.gov/topic/Watersheds/basins.

49. Martin, *Physical Geography*, 151.

50. Thomas Madigan, "The Geology of the MNRRA Corridor," chap. 1 in *River of History: A Historic Resources Study of the Mississippi National River and Recreation Area*, ed. John O. Anfinson (St. Paul: Corps of Engineers, 2003), 27.

51. Dott and Attig, *Roadside Geology*, 180.

52. Carson and Rawling, "Late Cenozoic Evolution," 5–9.

53. Eric C. Carson, J. Elmo Rawling III, John W. Attig, and Benjamin R. Bates, "Late Cenozoic Evolution of the Upper Mississippi River, Stream Piracy, and Reorganization of North American Mid-Continent Drainage Systems," *GSA Today* 28, no. 7 (January 19, 2018), 6, www.geosociety.org/gsatoday/science/G355A/article.htm.

54. For more details on this story, see Spoolman, *Wisconsin State Parks*, 72–79.

55. Dott and Attig, *Roadside Geology*, 166.

56. Martin, *Physical Geography*, 152.

57. Madigan, "MNRRA Corridor," 21.

58. Martin, *Physical Geography*, 165–66.

59. Madigan, "MNRRA Corridor," 34. A good topographic map shows clearly how Lake Pepin is being filled; see DeLorme Map-

ping Co., *Wisconsin Atlas & Gazetteer*, 14th ed. (Yarmouth, ME: Garmin Ltd., 2017), 52.

60. P. G. Richerson, R. Boyd, and R. L. Bettinger, "Was Agriculture Impossible during the Pleistocene but Mandatory during the Holocene? A Climate Change Hypothesis," *American Antiquity* 66, no. 3 (2001): 387–411.

61. Owen Jarus, "Cahokia: North America's First City," *Live Science* (January 12, 2018), www.livescience.com/22737-cahokia.html.

62. D. Klinkenberg, "Mississippi River History and Travel," https://mississippivalleytraveler.com (Klinkenberg is the author of a series of Mississippi River guidebooks); and

Thomas F. Waters, *The Streams and Rivers of Minnesota* (Minneapolis: University of Minnesota Press, 1977), 219.

63. Dott and Attig, *Roadside Geology*, 165.

64. US Fish and Wildlife Service brochure for Upper Mississippi National Fish and Wildlife Refuge, dated August 2014, www.fws.gov/uploadedFiles/UMR2011(3).pdf.

65. Dott and Attig, *Roadside Geology*, 178.

66. Dott and Attig, 175.

67. For more details and trail guides, see Spoolman, *Wisconsin State Parks*, 72–79.

68. Carson and Rawling, "Late Cenozoic Evolution," 18.

69. For more details and trail guides, see Spoolman, *Wisconsin State Parks*, 99–105.

Selected Bibliography

Bogue, Margaret B. *Around the Shores of Lake Superior: A Guide to Historic Sites.* Madison: University of Wisconsin Press, 2004.

Carson, Eric C., and J. Elmo Rawling III. *Late Cenozoic Evolution of the Lower Wisconsin River Valley: Evidence for the Reversal of the River.* Madison: Wisconsin Geological and Natural History Survey, 2015.

Carson, Eric C., J. Elmo Rawling III, John W. Attig, and Benjamin R. Bates, "Late Cenozoic Evolution of the Upper Mississippi River, Stream Piracy, and Reorganization of North American Mid-Continent Drainage Systems," *GSA Today* 28, no. 7 (2018): 4–11.

Cordua, William S. *Geology of the St. Croix Valley, Wisconsin and Minnesota: Guidebook for 10th Annual UW Geologic Field Conference.* River Falls, WI: n.p., 1978.

Cornell, Richard D. *The Chippewa: Biography of a Wisconsin Waterway.* Madison: Wisconsin Historical Society Press, 2017.

Deller, Howard, and Paul Stoelting. "Wisconsin's Door Peninsula and Its Geomorphology." *The Wisconsin Geographer* 2 (1986): 29–41.

Derleth, August. *The Wisconsin: River of a Thousand Isles.* Madison: University of Wisconsin Press, 1942.

Dott, Robert H., Jr., and John W. Attig. *Roadside Geology of Wisconsin.* Missoula, MT: Mountain Press Publishing, 2004.

Durbin, Richard D. *The Wisconsin River: An Odyssey through Time and Space.* Cross Plains, WI: Spring Freshet Press, 1997.

Ecological Landscapes of Wisconsin Handbook. Madison: Wisconsin Department of Natural Resources, 2011.

Gard, Robert. *The Romance of Wisconsin Place Names.* 2nd ed. Madison: Wisconsin Historical Society Press, 2015.

Green, Doris, with Michael H. Knight. *Explore Wisconsin Rivers.* Madison: Trails Books, 2008.

Greenberg, J. K., B. A. Brown, L. G. Medaris, and J. L. Anderson. "The Wolf River Batholith and Baraboo Interval in Central Wisconsin." *32nd Annual Meeting of the Institute on Lake Superior Geology.* Madison: Wisconsin Geological and Natural History Survey, 1986.

Hedquist, David. *Waterfalling in Wisconsin: The Complete Guide to Waterfalls in the Badger State.* Boulder: Trails Books, 2014.

Kluessendorf, Joanne, and Donald G. Mikulic. "The Lake and the Ledge: Geological Links between the Niagara Escarpment and Lake Winnebago." *65th Annual Tri-State*

Geological Field Conference Handbook, Menasha, WI: Weis Earth Science Museum, 2004.

Koenigs, Ryan, Ronald Bruch, Donald Reiter, and Dave Grignon. "Lake Sturgeon Reintroduction and Movement in the Upper Wolf River, Wisconsin." *American Fisheries Society 144th Annual Meeting*, Quebec City, Canada, August 17–21, 2014.

LaBerge, Gene L., *Geology of the Lake Superior Region*. Tucson: Geoscience Press, 2004.

Lange, Kenneth I. *Song of Place: A Natural History of the Baraboo Hills*. Baraboo, WI: Ballindalloch Press, 2014.

Leopold, Aldo. *A Sand County Almanac*. Oxford: Oxford University Press, 1949.

Loew, Patty. *Indian Nations of Wisconsin*. Rev. 2nd ed. Madison: Wisconsin Historical Society Press, 2013.

Luczaj, John A. "Geology of the Niagara Escarpment in Wisconsin." *Geoscience Wisconsin* 22, no. 1 (2013).

Martin, Lawrence. *The Physical Geography of Wisconsin*. Madison: University of Wisconsin Press, 1965.

McPhee, John. *Annals of the Former World*. New York: Farrar, Straus and Giroux, 1981.

Mickelson, David M., and Susan L. Hunt. *Landscapes of Dane County, Wisconsin*. Madison: Wisconsin Geological and Natural History Survey, 2007.

Mickelson, David M., Louis J. Maher Jr., and Susan L. Simpson. *Geology of the Ice Age National Scenic Trail*. Madison: University of Wisconsin Press, 2011.

Mikulic, D. G., and J. Kluessendorf. *Wauwatosa's Ancient Reef and Amateur Naturalist: Dedication of the Schoonmaker Reef and the Fisk Holbrook Day Home as National Historic Landmarks*. Wauwatosa, WI: Wauwatosa Historical Society, 1998.

Miller, G. Tyler, and Scott Spoolman. *Living in the Environment*. 20th ed. Boston: Cengage Learning, 2021.

Outwater, Alice. *Water: A Natural History*. New York: Basic Books, 1996.

Paull, Rachel K., and Richard A. Paull. *Wisconsin and Upper Michigan*. K/H Geology Field Guide Series. Dubuque: Kendall/Hunt Publishing Company, 1980.

Quimby, George I. *Indian Life in the Upper Great Lakes, 11,000 BC to AD 1800*. Chicago: University of Chicago Press, 1960.

Schultz, Gwen M. *Wisconsin's Foundations*. Madison: University of Wisconsin Press, 2004.

Shrock, Robert R. "Geology of Washington Island and Its Neighbors, Door County, Wisconsin." *Transactions of the Wisconsin Academy of Sciences, Arts and Letters* 32 (1940): 199–227.

Spoolman, Scott. *Wisconsin State Parks: Extraordinary Stories of Geology and Natural History*. Madison: Wisconsin Historical Society Press, 2018.

Stein, Carol A., Seth Stein, Reese Elling, G. Randy Keller, and Jonas Kley. "Is the Grenville Front in the Central United States Really the Midcontinent Rift?" *GSA Today* 28, no. 5 (2018).

Thorson, Robert. *Beyond Walton: The Geology, Ecology, and Cultural History of Kettle Lakes from Maine to Montana*. New York: Walker & Company, 2009.

Travis, Jack W. "Geology of Brown and Door Counties, Wisconsin." In *Field Trip Guide Book: Geology of Brown and Door Counties, Wisconsin, May 30–31, 2009*, ed. American Institute of Professional Geologists. Westminster, CO: AIPG, 2009.

Treuer, Anton. *Atlas of Indian Nations*. Washington, DC: National Geographic, 2014.

Waller, Donald M., and Thomas P. Rooney, eds. *The Vanishing Present: Wisconsin's Changing Lands, Waters, and Wildlife*. Chicago: University of Chicago Press, 2008.

Waters, Thomas F. *The Streams and Rivers of Minnesota*. Minneapolis: University of Minnesota Press, 1977.

Wisherd, Nan. *Brule River Country*. Brule, WI: Cable Publishing Company, 2017.

Index

Note: Page numbers in *italics* refer to illustrations.

About the Author

Scott Spoolman is the author of the Wisconsin Historical Society Press book *Wisconsin State Parks: Extraordinary Stories of Geology and Natural History*. As a writer, he has focused on the environmental sciences, especially those stories of science and the environment related to Wisconsin and surrounding states. After earning a master's degree from the University of Minnesota School of Journalism, he worked for several years as an editor in the education sector, specializing in textbooks and other educational materials. Since 1996, he has worked as a freelance writer and editor for a variety of outlets and has coauthored several editions of a series of environmental science textbooks. Throughout his life he has enjoyed exploring the forests and waters of the upper Midwest, as well as the mountains of the western and eastern United States. His passion for the outdoors led him to develop an avid interest in geology and natural history, especially in his home state of Wisconsin.

GAIL MARTINELLI